D0402360

DARK
NATURE

ALSO BY LYALL WATSON

Dreams of Dragons
Gifts of Unknown Things
Heaven's Breath
Lifetide
Lightning Bird
The Nature of Things
The Romeo Error
Supernature
Supernature II
Whales of the World

DARK NATURE

A Natural History of Evil

LYALL WATSON

HarperPerennial
A Division of HarperCollins*Publishers*

A hardcover edition of this book was published in 1996 by HarperCollins Publishers.

HarperCollins books may be purchased for educational, business, or sales promotional use. For information please write: Special Markets Department, HarperCollins Publishers, Inc., 10 East 53rd Street, New York, NY 10022.

First HarperPerennial edition published 1997.

Designed by R. Caitlin Daniels

The Library of Congress has catalogued the hardcover edition as follows:

Watson, Lyall.
Dark nature : a natural history of evil / Lyall Watson. — 1st U.S. ed.
p. cm.
Includes bibliographical references and index.
ISBN 0-06-017688-1
1. Good and evil. 2. Philosophy of nature. 3. Man. 4. Biology—Philosophy. 5. Evolution. I. Title.
BJ1406.W37 1996
111'.84—dc20 96-1663

ISBN 0-06-092790-9 (pbk.)

97 98 99 00 01 ❖/RRD 10 9 8 7 6 5 4 3 2 1

By the pricking of my thumbs,
Something wicked this way comes.
Open, locks, whoever knocks.

Macbeth IV i
William Shakespeare

CONTENTS

INTRODUCTION
By the pricking of my thumbs . . . / ix

PART ONE—DARK NATURE / 1

ONE
Being Good: The Ecology of Evil / 11

TWO
Breaking the Rules: The Arithmetic of Evil / 48

PART TWO—HUMAN NATURE / 89

THREE
Being Bad: The Ethology of Evil / 104

FOUR
The Evil That Men Do: The Anthropology of Evil / 141

FIVE
The Wages of Sin: The Psychology of Evil / 182

PART THREE—EVIL NATURE / 233

SIX
The Mark of Cain: The Identity of Evil / 249

BIBLIOGRAPHY / 293

INDEX / 309

INTRODUCTION

By the pricking of my thumbs . . .

I live on a boat, on an ocean-going trawler, and find that I think best when I can do so alone, offshore.

I am an old-fashioned field naturalist, a beachcomber at heart, with a passion for bones and stones and seashells, prone to letting delight in such detail saturate my whole attention. I become easily enrapt and, in that happy state, lose sight of the larger environment. It ceases to environ me. It becomes an extension of me. And I cannot smell myself.

So every once in a while, I have to fall back and look for blue water, floating out to where the horizon is flat and I can find some degree of alienation. Just enough to get a glimpse again of the whole ecology, of the form and process of things.

The best position is that of the offshore islander, of men like Yeats and Joyce and Shaw, who could see the outline of the distant mainland and yet still enjoy sufficient detachment to permit pattern recognition.

There is no better think tank in my experience than the tropical blue of the great Gulf Stream, where the ocean has a desert aspect, where birds become scarce and even flying fish break the surface infrequently. Where the landscape consists largely of scattered cumulus clouds, through which one passes effortlessly, carried along by a river of deep warm water flowing ceaselessly between invisible banks.

It was here that the idea found me, courtesy of an errant broadcast that slipped from its assigned frequency into the window of a radio watch I keep in case of maritime emergencies. The voice was a familiar one, that of another islander halfway round the world, of Arthur C. Clarke on his beachhead in Sri Lanka, giving yet another interview through the medium of a satellite system whose existence he predicted twenty years before it was launched in 1965.

He was being asked about the possibility of extraterrestrial intelligence and was replying, as usual, with that mixture of optimism and longing that makes his science fiction so attractive. But the interviewer was in no mood for metaphysics and kept pushing for sensation, asking, "Where are these other civilizations?" and "When are we going to meet them?"

Clarke responded in his usual fluid way, refusing to be drawn into quick and easy answers, but then the interviewer changed direction and provoked a reply I had never heard before. "If we do eventually meet them," she asked, "should we be afraid?"

Clarke was obviously intrigued by this approach.

"It is possible," he said, "that alien races could be warlike and aggressive. Although we used to find solace in the notion that if they were truly evil, they would destroy themselves long before they ever got the chance to become a threat to us . . . "

Then there was a long pause before the master added quietly, ". . . but I wouldn't count on it."

Indeed. It would be foolish to ignore the possibility that evil is not peculiar to human nature. There seems to be a lot of it about right now, and maybe it is not confined to our particular ecology at all. It could even be universal. And if it is, then it becomes necessary to see evil as a force of nature, as a biological reality. As something that could appear in any environment, no matter how alien, and exercise an evolutionary influence, here or there, having separate survival value and its own strange agenda.

Clarke's throwaway line started a whole train of thought for me out there somewhere west of Bimini. It opened doors I had not dreamed of looking through before—and this is the result.

A natural history of evil. A look at the origin and the meaning of being wicked. A study of the dark side of nature, made in biological terms because that is what I do, and because I think it time that evil was examined in ways other than those already well traveled by religion, moral philosophy and criminology. I believe each of these approaches fails in the end, not because they underestimate evil, but because they misunderstand its nature.

Milton, Dante, Goethe and Robert Louis Stevenson come, I think, closer to the truth and I have profited from their insights, but this is not a work of fiction. It is a very individual view of nature, one taken by a curious naturalist afflicted with pricking thumbs, drawing on his own experience—and on recent findings—in evolutionary biology, anthropology and psychology—to underpin some of our oldest intuitions.

My concern begins with the perplexity we all share when faced with stories such as that of a ninety-two-year-old widow, deaf and partially sighted, who has her wrist and hip shattered when she is knocked to the ground by a young assailant whose sole aim, it seems, was to steal her white stick.[369*] With the anguish we feel when confronted by the behavior of two otherwise unremarkable ten-year-old boys who kidnap and torture a baby before beating him to death beside a railway line.[211] And with our total incomprehension when a law student who has savagely raped and murdered at least twenty-eight attractive young women, leaving his bite marks on their mutilated bodies, has dozens more women exactly like his victims blushing and giggling on the front benches of the courtroom during his trial.[329]

We are in murky territory here. And navigation through it is made no easier by labeling such behavior as "aberrant." It is obviously unusual, but it may not be unnatural. We need to know a great deal more about its biological roots and its organic history before dismissing it as "psychopathic" and therefore beyond ordinary understanding. I believe it will help to know that evil is commonplace and widespread, perhaps not even confined to our species. But at the same time we ought to be aware of the dangers

*Numbers refer to sources listed in Bibliography.

of misinterpretation. It is all too easy to leap to unwarranted conclusions, particularly where other species are concerned.

I have stood on a windswept beach in Patagonia, where the deserts of southern Argentina run down to meet cold waters off the Valdez Peninsula, and found myself faced with just such an unexpected, but not untypical, moral dilemma. I was there to watch large baleen whales which come in unusually close to that shore to calf. We know them as "right" whales because they are slow and easy to hunt and have bodies so rich in oil that they float even when dead. These days, happily, such whales are no longer hunted except by a growing band of whale-watchers. But my visit was made late in the season, right whales were scarce and I found myself more interested in the behavior of "wrong" whales, the ones we have given a bad name and called "killer" or *orca,* from the Latin for a demon or denizen of hell.

Killer whales are predators, catching and eating sharks, rays, otters, walrus and even other whales. In Patagonia, they have developed a taste for sea lions, each consuming as many as three pups a day when these are available, taking them from the shallows or plucking them directly from the beach after an awesome charge which carries the five-ton whales high up the sloping shore, right in amongst sunbathing sea lions.

On this occasion, two male whales, each having eaten their fill, were simply playing with their food. The last item on their menu, a live and terrified sea lion pup, was being tossed from one to the other, flying through the air between them like a beach ball. Each time the sea lion tried to set off back to the beach, struggling to reorient itself after splashdown, one of the players would go off after it with every evidence of excitement, putting the pup back into play with an athletic flourish.

It was easy, watching this performance, to feel compassion for the hapless pup and to see the adult whales as cruel killers. Even for a biologist used to the cut and thrust of predatory behavior, familiar with the need for hunters to practice and perfect the moves of their profession, it wasn't easy to watch without wincing. My discomfort was heightened by my knowledge of the intelligence and creativity of these particular predators, which made it difficult for me to cut them the sort of moral slack one allows a kitten found playing with a mouse. And I was finding it easy instead to feel censorious and to believe in the biological roots of evil and the inevitability of whales "acting like animals," when the killers themselves confounded me.

The larger of the two whales, whose fin was so tall it curved over at the tip, the very one who seemed to take most pleasure in the game, brought it to an astonishing end. He cut short the play as abruptly as if "time-out" had been called, picked up the sea lion in his huge curved teeth and carried it toward the shore. In deep water close to the beach he put on a burst of speed that carried him on the crest of a wave high up above the waterline, where he replaced the pup almost where it had been found, nudged the motionless little sea lion until it suddenly sprang to life and floundered, ruffled but otherwise apparently unharmed, back to its family. Then the whale seemed to turn directly to me, holding a look long enough to make me feel uncomfortable with my presumptions, even out there in my element, before he heaved back into his own.

There are obviously no easy answers to the problem of evil, but I have become convinced that there must be more constructive, more natural and relevant ways of describing and defining it. Part of the problem is simply one of perspective, of the way in which we have become accustomed to thinking about our relationship to the rest of nature.

We like to see ourselves as something more than animals. As animals *plus.* The plus factor being some kind of special human essence that has been added to the baseline of animality. This additive has been defined in a variety of ways, ranging from culture to consciousness, from religion to morality, but the truth is that the mystery ingredient, if it exists at all, remains just that, a mystery.[212]

The problem with this approach is that it looks at life from the top down. It excludes humans, and human nature, from the rest of nature. And it relegates the rest, plants and animals alike, to some kind of mindless substrate.

The world does not look like that to me. As a naturalist, I tend to see things from the bottom up. I define my science as that which is concerned with life in all its guises, including humanity at the upper end of a spectrum of complexity. At every level on that scale are individual organisms that, even at their most simple, actively select and change their environments, and are themselves changed in the process. And I find, along with other biologists concerned with behavior, that principles derived from simple organisms can very often help us to understand ourselves.[186]

We know, for instance, that it no longer makes sense to see behavior either as innate or acquired, inborn or learned. Some patterns, such as the human smile, are universal. All humans do it, even babies born blind and deaf, and we all use the same set of facial muscles. But the circumstances in which smiling takes place, and the significance attached to it, vary enormously.[116] That much we learn.

We know too that each species is constrained by what it *can* learn and has a tendency to learn some things rather than others. In every hedgerow in Britain, one of 7 million nesting pairs of chaffinch cements its bond with a characteristic song.[184] The pink-breasted male bird announces his residence, and his reproductive intentions, with three delicate phrases ending in an artistic flourish. But he has to hear this song to sing it. No chaffinch brought up in isolation sings it, and no chaffinch anywhere ever learns any other song. The music is written somewhere in chaffinch chemistry, but it rests there, mute and inglorious, until summoned up by the performance of another male in full throat.[364]

We are similarly constrained. Widespread human phobias, for instance, the sort that seem to be independent of culture and experience, all touch on historical hazards. They appear in response to darkness, falling, snakes, spiders or being alone. The sort of things with which we had to deal earlier in our evolution, rather than those things which most often kill people now—such as automobiles, hypodermics and handguns.[185]

We are also, like the chaffinch, predisposed to learn certain kinds of things. We are hard-wired for sucking, chewing and walking on two feet. And, far more mysteriously, without conscious effort or formal instruction, we spontaneously develop the complex skills involved in the acquisition and use of a spoken language. We have an instinct for words, spinning them as surely and as naturally as a spider spins its web. They are part of our birthright, a recipe for community.[304]

As a biologist, I was taught two Darwinian truths. First, that every living thing is a unique historical entity, distinct from its predecessors and its environment. And second, that it came to be that way as a result of variability introduced by reproduction. I am content still to accept both tenets. They help to explain how natural selection takes place and why it leads to evolution. They deal well with events, but they do not deal at all successfully with the process, with the interactive and never-resting stream of events that produce the flow we know as "life."

Life is not something added to matter to make it move, any more than culture is added to animality to make it human. Life is our name for the process going on in an evolving organic world. It describes the interactions taking place in any ecology. Organisms are alive, but life is not something "in" them. Organisms are "in" life. They grow there, change there, and by being there, change the nature of their surroundings.

The last point is important. Much thinking about evolution makes the implicit assumption that an environment is something given, something that has to be there from the start, to which organisms need to adapt. Ecological niches are seen as existing before the organisms that come to occupy them, and environments are described as sets of constraints to which groups of genes are forced to respond. But things, in my experience of them, seldom work like that. There appears instead to be a more important unity at work, a dynamic flux which involves both organism and environment, linking them in an inner way, allowing the combination to select, from a wide range of possible variations, the circumstances and the forms that actually appear.

In history we make ourselves, creating from within us the world in which we are involved. And the same, I suspect, is true of all species. They make themselves, contributing to the whole history of life. This is my kind of biology. The study of organisms that are not just manufactured by and for the benefit of their genes, but have their own ideas. They come into being as the result of genetic expressions, but only insofar as these can unfold in sympathy with the field in which they are found. It all depends, in the end, on who reads the instructions in the genetic code, and on where these instructions are carried out.

Mine is a biology flexible enough not to need to define humans as animals-plus-anything. It is a creed comfortable with the notion that we, and all other organisms, are involved in what the philosopher Alfred North Whitehead described as "a creative advance into novelty."[413]

This is the position on which I stand, the point from which I make my forays into nature. Everything I do is informed by the basic assumption that life, even if it fails to follow any grand design, nevertheless makes sense on its own terms. It has meaning and the pieces of the puzzle tend to fit together, often astonishingly well. And I see no reason to exclude evil from this scheme of things.

Evil exists and seems to me to have sufficient substance to give it credence as a force in nature and as a factor in our lives. It is part of the ecology and needs to be seen as such. My thumbs convince me, not that "something wicked this way comes," but that it is already here and has been for a very long time, casting its shadow on almost everything we do.

I know it is not easy to deal with evil, but it is very necessary that we should confront it. We need to take a long hard look at wickedness, wherever it occurs and however uncomfortable it might make us feel. It is far, far more dangerous to deny it and to look the other way.

So here goes. "Open, locks, whoever knocks . . . "

LYALL WATSON: On the *Amazon*

PART ONE

DARK NATURE

NO ONE BECOMES DEPRAVED ALL AT ONCE.
Juvenal

Evil has a head start. It compels our attention. Villains grip the imagination with a force no hero can hope to match. Who can remember the compass point from which the Good Witch came? And who can forget Rasputin or Torquemada, Hitler or Stalin, Jack the Ripper or Hannibal Lecter?

Evil seems self-evident. It exists, doesn't it? There are evil acts and evil individuals, aren't there? Torture is undeniably wicked, surely, whatever its motives, even if the *auto da fé* and all the other cruelties of the Inquisition were prompted by concern about witchcraft and heresy. Murder must be evil and denounced wherever it is found, mustn't it? But what of killing designed to speed a painless death or to avoid needless suffering? Is that wickedness or kindness? And how do we classify killing authorized, even demanded of us, by the state? Are there good wars? Does it depend on who you are asked to fight for, or against?

Our ambivalence about all these things is highlighted by contradictions contained within the very words we use to describe or explain them. The English word "sanction," for instance, is taken to mean the giving of permission or authority to an action by law or custom. And yet it is applied with equal vigor and conviction to measures designed to prohibit such actions. It is both punishment and reward, and seems to acquire this slippery status as a direct result of its derivation. "Sanction" is adopted from the Latin *sanctio*—"to make sacred." And anything goes, it seems, once something can be seen to be in the interest of a god.

The only certainty about evil is that it is difficult to pin down and needs careful definition. So I want, at the outset, to set out my terms of reference as clearly as I can.

I will refer to, sometimes even defer to, classical concerns, but without getting involved in any intellectual debate about free will or human destiny. I will also try to avoid religious obsessions with sin and the enduring paradox of good gods who allow bad things to happen. The literature in both these areas is extensive and yet doesn't seem to me to come to any useful conclusion, at least of the kind which might help to shed light on the origin and evolution of evil.

I will draw, sparingly, on another growing resource: the published reports of "true crimes," which are rich in bloodcurdling detail, but are not particularly revealing about the minds and motives of multiple murderers such as Myra Hindley, Dennis Nilsen and Jeffrey Dahmer. The lives and activities of these specialists in certain kinds of evil exercise the kind of fascination for us that is appropriate to a road accident, and should perhaps be the subject of equally transient attention.

What I am concerned to do is to track down and bring together and reevaluate all those ongoing, everyday pieces of natural history and human behavior that might contribute to a more biological view of evil as a force of nature. My source material lies largely in the public domain. It surfaces in those aspects of animal behavior that have to do with sex, territoriality and aggression. It is evident in courtship, combat and other quarrels about limited resources in any species. It occurs as a constant refrain in human matters of love, hate, jealousy and greed. Its faces are equally familiar to beach-walkers or bird-watchers, to those whose fortune it is to raise families, or those whose misfortune it is to wage war. It is the focus of much of the most compelling ritual, tradition, mythology and folklore. And concerns about it seem still to lie at the heart of all the world's great faiths.

Nothing is necessary to study such things but curiosity and a willingness to concede that nature has a dark side that deserves as much attention as we already lavish on wildflowers, butterflies, rainbows and sunsets. And to set some clear constraints on the area of my inquiry, I choose to begin in the time-honored way with a working definition of evil as suggested by the *Oxford English Dictionary.*

Evil, together with the German *übel* and the Dutch *euvel,* derives from the ancient Teutonic form *ubiloz.* This refers largely to things which are the opposite of "good." The dictionary points out that in modern usage "evil" tends to be superseded by some softer word such as "bad," and that this may even, in modern slang, have come to mean "good." "Evil" is clearly still common in English in its substantive form, describing occasions involving harm, injury, disease or misfortune. But, in the opinion of the dictionary's editors, the adjectival form of "evil," in the old sense meaning "wicked," may now be almost obsolete as applied to persons.

That may have been true in 1989, but events in the five years which have elapsed since the publication of the second edition of the dictionary make it necessary to reconsider the status of active evil as a force in modern society.

Nancy Gibbs, reporting for *Time* magazine in 1994 on the death of half a million people in the latest tribal massacre in Africa, struggled for comprehension and found herself forced to fall back on the traditional assessment of a missionary. "There are no devils left in Hell," she said. "They are all in Rwanda."[140]

"Actually they brought Hell with them," wrote Gibbs. "You only have to watch the rivers for proof. Normally in this season, when the rains come to these lush valleys, the rivers swell with a rich red soil. They are more swollen than ever this year. First come the corpses of men and older boys, slain trying to protect their sisters and mothers. Then come the women and girls, flushed out from their hiding places and cut down. Last are the babies, who may bear no wounds: they are tossed alive into the water, to drown on their way downstream. The bodies, or pieces of them, glide by for half an hour or so, the time it takes to wipe out a community, carry the victims to the banks and dump them in. Then the waters run clear for a while, until men and older boys drift into view again, then women, then babies, reuniting in the shallows as the river becomes their grave."

Analysts of the situation point to suppressed tribal rivalries, the rupture of natural authority, and other legacies of colonial rule, but African readers have been quick to take more personal responsibility for the chaos and the shame. "There is no excuse," they said. "It was we who set the terrain for this devil's carnival."[380] The consensus seems to be that such conflicts are likely to haunt our world more and more often in the coming decades as we return, for a number of reasons, to an almost medieval situation in which wickedness is real, raw and evident.

One feature which does emerge from the dense columns of the *Oxford English Dictionary* entry on evil is an interesting taxonomy. It seems clear that there are "strong" and "weak" meanings of the word. The strong sense of evil is the active one which means "morally depraved," "truly bad" or "thoroughly wicked." The antithesis of good in all the principal senses of that term. The weak version describes that which is merely "unpleasant," "harmful," "disagreeable" or "likely to cause discomfort."

This classification of evil, which mirrors the division of nuclear "weak" and "strong" forces in nature, is a useful one to which I will return. For the moment, it is sufficient to note that the root form of *ubiloz* actually means "up" or "over," and implies a primary sense which the dictionary describes as "going beyond due measure" or "overstepping the proper limits." Something excessive.

That too feels right and useful, because even in cultures which have no actual word for evil, there is disquiet and concern about influences which destroy integrity or disrupt equilibrium. A person becomes bad by reason of their pursuit of such excess. The object of the pursuit is not evil of itself, but may be seen to become so if the desire for it interferes with the well-being of oneself or others.

This universal understanding was refined into a coherent philosophy over two millennia ago by another beachcombing naturalist during twelve formative years he spent in exile on the islands of the Aegean. By the time he returned to Athens in 336 B.C. to found the first Lyceum, Aristotle was ready to write and to teach the first treatise on ethics as a practical science.

Aristotle, notwithstanding Francis Bacon's rejection of him in the seventeenth century, was the first truly empirical scientist. Unlike Bacon, who never even touched a test tube, Aristotle was a brilliant and meticulous observer, experimenting and dissecting, testing every hypothesis as he went along, learning and teaching on the job. He was, in bold contrast to much of prevailing Greek practice, a hands-on, field naturalist, believing that knowledge of the real world could only be acquired by going out and looking at it, instead of just thinking about it. He made mistakes, but these are obvious only in hindsight, with developments such as quantum theory and the electron microscope. His biology, all of it based on detailed observation of nature, has stood the test of time. His philosophy and creative logic remain the bedrock of speculative thought. And his discussions on morality and virtue set ethics on the same practical level as politics; the object of both disciplines being not only to know the difference between right and wrong, but to act on such knowledge.

The *Nicomachean Ethics,* named for Aristotle's son, is not an easy read.[15] It meanders along with the cadence of the spoken rather than the written word, which is perhaps not surprising as it seems to have survived only in the form of lecture notes transcribed by students direct from their master's peripatetic instruction. But the drift is clear and the text shines with the sort of down-to-earth, outdoor common sense one would expect from a thoughtful naturalist.

"Desire" is the key to Aristotle's moral philosophy. He recognizes "right" and "wrong" desires, and suggests that one can be forgiven for a wrong action—we all make mistakes—but never for a wrong desire. His list of such moral evils includes greed, lust, avarice, sloth, anger, envy and pride. Which nicely covers all of the Seven Deadly Sins. It is instructive, in this respect, to compare Aristotle's blacklist with the celebrated Commandments of Judeo-Christianity. Only one of those ten proscriptions, that of covetousness, is a wrong desire. All the rest deal with actions. They are the rules of a society more concerned with what one should or should not do than with understanding the motives behind murder, theft, adultery and perjury.

Aristotle wanted to know why. His approach to the world was very modern, in the sense that science since his time has struggled to sort out those parts of the world which seem, in the mass, to be indistinguishable one from the other. His was a tidy mind and it insisted on taxonomies, even of desire. In addition to right and wrong desires, he recognized "acquired" desires, which differ from individual to individual, and "natural" desires, which are a part of human nature. These are what most of us now would recognize simply as "wants" and "needs."[2] Needs are the sort of desires that are inevitable and necessarily good. Needs are never wrong. Wants are more problematic and Aristotle assessed them by their quantity. Good wants, he said, are right desires as long as they are identical with our needs for them. They become bad, and may be identified as wrong desires, if we want too much. But—and this I think is Aristotle's most vital contribution to the discussion of evil, contradicting the schools of hair-shirt discipline which turn abstinence into a virtue—such desires may be bad and wrong also—if we want *too little.*

Aristotelian ethics is the ethics of "just enough." Neither too much nor too little. Enough is enough, even of a good thing. Any more or less falls outside what he called the "golden mean" and fails to contribute to the whole good, the *totum bonum.* Even moral virtues such as courage are good only if they lie along the narrow mean. A man who fears everything becomes a coward, but the one who fears nothing is a dangerous fool.

Looked at in this way, good and evil are not a matter of taste or fashion, like or dislike. They are ideas rooted in our tissues. They are forces subject to natural law and should be assessed only on their effect, on what they add to or subtract from what the Greeks called *eudaimonia.* This is usually translated as "happiness," but that fails to convey the whole meaning of something that includes an active component, a sense of movement toward a goal. "Wellbeing" is perhaps closer to what Aristotle had in mind as the aim of a good life. Reading between his lines, it looks less like "survival of the fittest" and far more like "the fitting of as many as possible to survive."

What strikes me most forcibly is Aristotle's feeling for ecology, 2,000 years and more before that science came into being. He even recognized the peculiar human dilemma of having to distinguish between living and living well, suggesting that right desires had to do with ultimate rather than immediate ends. And that no act or desire could be judged in process, only once it was complete. In a wonderful flourish which long anticipated Thomas Jefferson, Aristotle made it clear that we have no basic natural right to happiness, only to "the pursuit of happiness." He concluded that any society which made it impossible for all its citizens to pursue such a goal was necessarily bad. And that any individual who consciously or deliberately deprived another individual of such opportunity was unquestionably evil.

This is a very bald summary of Aristotelian ethics, but it does help to provide a framework for my exploration of the evolution of evil. If "good" can be defined as that which encourages the integrity of the whole, then "evil" becomes anything which disturbs or disrupts such completeness. Anything unruly or over the top. Anything, in short, that is bad for the ecology.

ONE

BEING GOOD: THE ECOLOGY OF EVIL

We live in a designer universe. Things in it have a glorious and unsettling tendency to be beautiful, and it is not immediately obvious why that should be so.

If evolution depends on chance alone, then there would seem to be equal opportunity for awkwardness and ugliness. But nature very seldom turns out like that. Even its least enduring designs are intriguing, and the successful ones take your breath away. Most often they do this by their simplicity. By the way they solve complex environmental problems with economy and austerity, producing solutions that not only work well, but are also aesthetically satisfying.

I spend a lot of my time on low-tide lines, where the long rhythms of land and sea combine. It is an area of unrest, an elusive boundary along which creation still occurs. Every day spent there is different and on most of them one comes face to face with miracles of design.

I find solace in seashells, as humans always have, transforming them into ritual objects, symbols and metaphors for a thousand other things. There is something about the architecture of a shell that encourages such association, conjuring up images of fortresses and graves, of death and resurrection. And none is more evocative than the shell of the chambered nautilus, a living fossil which drifts along near the floor of the Pacific Ocean, successfully eluding extinction for over 500 million years, moving from chamber to chamber as it grows. The animal involved is a kind of squid, mottled white and brown, with ninety-four small tentacles and a natural flair for applied geometry. Its mantle cells secrete a limestone shelter with a pattern to match its own, gently bent to fit the bodyline. And as it grows, it adds on progressively larger chambers along a curve that forms a perfect logarithmic spiral.

Spirals are the natural curve of life and uniform growth. They are always growing and yet never cover the same ground. They are the only form of curve in which one part differs from another in size, not in shape. Spirals work both ways, coming from and going to their source. They define and illuminate what has already happened, as well as leading inevitably on to bigger things and new discoveries. They are explanations of the past and prophecies of the future. And the best of them, including those drawn by the nautilus, conform to a ratio of exactly 1:1.618034, which alone among curved lines can be extended to infinity.[79]

This may seem like a heavy burden for a little deep-sea mollusk to bear, but it does so to perfection, prolonging the Ordovician through a dozen intervening geological periods right into the present and on to beaches everywhere in the Indopacific. By contrast, their relatives, the ammonites, are now totally extinct. Those coiled fossil forms are common everywhere in marine deposits from the Devonian to the Cretaceous, but the last one died more than 60 million years ago. And a clue to their demise lies perhaps in the fact that their spiral pattern was imperfect. Its sweep was a little flatter, not quite wide enough for truly vital expansion, not corresponding so nearly to the curve of perfect growth as the still-living nautiloids. It is not the good that die young, but those with bad designs.

I have never seen a live nautilus. They drift at depths beyond my reach, but I rejoice in the signs of their continued success, and check each outgrown shell I find on tropical tidelines, just to make sure they are keeping the mathematical faith and running true to form. So far so good.

Spirals seem to be inevitable. They are the perfect symbols of change and growth, of order within chaos, of opposites which are nevertheless the same. Small wonder that Jacob Bernoulli, the Swiss mathematician who pioneered the study of fluid dynamics, was also so fascinated by the mystery and magic of their symmetry that he had a spiral carved on his tombstone accompanied by the Latin inscription *Eadem mutata resurgo*—"Though changed, I shall arise the same."

He was the first to notice that water, air and even the galaxies themselves are strangely fond of spirals. Ours certainly is. The 100,000 million stars we call the Milky Way are grouped into a slowly rotating, disk-shaped assembly, each following its own orbit around the center in obedience to the simple laws of physics. The individual stars come and go, circling at different speeds, numerous enough to cancel each other out in a shapeless blur. By rights there should be no particular pattern to this rotation, no visible or persistent arrangement of stars within this arena. But there is. Our galaxy, like others in the neighborhood, is carefully choreographed. It has a stable and beautiful spiral structure, with arms of stars sweeping out in graceful curves like those of a Catherine wheel.

Such order is odd. A living giveaway. A definite sign that something untoward is happening, keeping things from winding down in the usual way. Our galaxy is far older than it should be. Everything else has a natural time cycle, a period during which it unwinds, but we have already passed our fail-by date and seem destined, according to the latest calculations, to survive for at least ten-to-the-power-of-forty times longer than we have the right to expect. Something is keeping us going, giving our galaxy, if not direction, then certainly an unexpected and unlikely form. One that just happens to be repeated in the life cycles of certain squid and cast up on remote beaches for the amusement of passersby.

A natural accident? I doubt it. The coincidence is too great to be attributed to chance. And the constant of 1.618034 is at least five decimal places too precise. I am not very good with numbers, but it does seem to me that they sometimes have a life of their own. An independence which makes it proper to speak of the discovery of, rather than the invention of, arithmetic.[277]

Amongst those who made the earliest discoveries was yet another beachcomber, an islander born in the Greek city-state of Syracuse and educated in Egypt. Archimedes did for mathematics what Aristotle, a century earlier, had done for biology, deducing law from his own experimental observations. He died that way in 212 B.C. during the Roman conquest of Sicily, stabbed by a soldier who fell upon him in the midst of calculation, drawing one of his famous figures in the sand.

Archimedes, fortunately, had already made a more permanent record of his thoughts, included amongst which is a treatise addressed to the King of Syracuse with the delightful title of *The Sand Reckoner.*[13] In it he argued against mystical claims to the effect that the number of grains of sand along the Sicilian seashore was beyond the power of man to count. Archimedes said he could do it. In fact he went a great deal further, setting up a new system of notation that he used to calculate the quantity of sand grains in the entire universe. And the number he came up with is astonishingly close to recent estimates of about 10^{40}.

Numbers this big are difficult to imagine, but ever since we began to do so, they have boggled the imagination even further by providing some astonishing coincidences. These are now gathered together and worried about under the provocative heading of The Lore of Large Numbers.[110]

There is no obvious reason why the age of the universe in natural atomic units should be related numerically to the number of particles in the universe, but there *is* a direct relationship. Both fall very close to 10^{40}.

Physics recognizes two natural time units. Hubble time, which is a way of measuring cosmological change; and nuclear time, which is the time taken for light to cross a single proton. There is no direct connection between the two, whose scales lie at opposite ends of the universal range, but the ratio between them happens to be 10^{40}.[99]

Studies of the strength of nuclear forces and the thermodynamic condition of the early universe fall within different and distinct branches of physics, and yet each produces the same constant of 10^{40}. The fine structure constants for gravity and electromagnetism would seem to be only accidentally connected, but once again each turns out to have a numeric value of 10^{40}.[314]

These are some of the large number coincidences. There are many more. So many that in 1938 the Nobel Prize–winning physicist Paul Dirac wrote: "Such coincidence we may presume is due to some deep connection in Nature." The coincidences are in fact so widespread that it is tempting to go much further than that, and to suggest that something very strange indeed is going on. Some sort of cosmic conspiracy.[195]

The British physicist Paul Davies talks about "cosmic cooperation of such a wildly improbable nature, it becomes hard to resist the impression that some basic principle is at work." He points out that recent discoveries about the early cosmos all suggest that it was set in motion with astonishing precision. Its nature not only encourages, but depends on, the coincidence of large numbers and a host of other numerical "accidents." Without these, stars like our Sun would not have come into being at all. "Had nature opted for even a slightly different set of numbers, the world would be a very different place. Probably we would not be here to see it."[99]

But we are, thanks to a long list of happy accidents, and an exquisite precision which British astrophysicist John Gribbin ascribes to what he calls "The Goldilocks Effect." Goldilocks, remember, was the nursery-tale heroine who was also perhaps the earliest recorded squatter. She moved in on the empty home of the Three Bears, trying out their beds, chairs and bowls of porridge until she found the ones that were neither too hard nor too soft, too big nor too small, too hot nor too cold—but "just right" for her. This rightness is an important idea, one with far-reaching consequences, not just for cosmology but for all life. And it has, I believe, direct relevance to the problem of good and evil—both of which may even have cosmic roots.

The way in which cosmic coincidence favors our existence is truly astonishing.

The fact that we find ourselves living on a solid surface, when the vast majority of the material in the universe takes the form of gas clouds or hot plasma, and the fact that we happen to be near a stable star, when many others are unstable or unable to support planets in any form, may not be coincidental. They may be inevitable. As far as we are concerned, it couldn't have happened in any other way. Life requires many strange circumstances, which perhaps arise only locally and for a comparatively short space of time. The chance of all the necessary conditions coming together in just the right way is infinitesimally small. But now that such conditions have occurred, and we are here, it is only fair to link our existence to that of the cosmos and to say, as John Wheeler, the professor of physics at Princeton, has said: "Here is man, so what must the universe be?"[411]

It may even be alive.

Life is strange. Whole disciplines have foundered in the attempt to define it. But even if we still have difficulty in deciding exactly what it is, there is one thing we may be certain it is not. It is never chaotic. Most other systems are. Left to themselves, they tend to become more and more disorderly, running down until their matter is randomly distributed. This is the natural state of things. Their future is written in the Second Law of Thermodynamics, which requires that entropy, the degree of disorder, increases until it reaches a maximum of complete chaos. This endpoint, the final result of the process of unwinding, is usually, and most revealingly, referred to as "the heat *death* of the universe."

Leaving aside, for the moment, the assumption implicit in this description that the universe is alive, we have at least a working definition of what it means to be so. Life is orderly. It brings order to randomness. It reverses the general trend by encouraging chance encounters. It thrives on unnatural events, bringing together ingredients which are wildly improbable in ways that are statistically unlikely. Life is a rare and unreasonable thing. It goes against the grain and can be recognized most easily by the fact that it is incongruous. It has form and order in the midst of surrounding confusion.

Life is something that leaps out at you, just as Earth itself does when seen rising over the stark horizon of a dead Moon in those memorable photographs taken by the Apollo astronauts. Compared to its neighbors in the solar system, our planet is a misfit with the soft blue look of an oasis in the desert. It is obvious, even from a distance, that something unusual is going on here, something cool and lively under the protective bubble of vapor.

Our planet breaks all the rules. In theory, and in accordance with the law, Earth ought by now to be in a state of barren disorder. Five thousand million years of existence should have been more than long enough to reduce our hot young world to a tired old one on which everything has leveled out and settled down. By rights, all energetic reactions should have gone to completion, all random and willful disturbance reduced to predictable uniformity. This third planet of an unremarkable star ought by now to be in a sober, steady state, null and void and lifeless. But, of course, it isn't.

Our world hums with creative energy, glorying in improbable assemblies of unlikely chemicals. Anarchy is loose in the land and succeeds, somehow, in countering chaos. And the result, against all the odds, is order. The ocean heaves and turns, rain falls, land rises, winds blow and life blossoms—all because something keeps our environment stable despite the vagaries of Sun and time. It looks astonishingly well organized, more like an artifact than an accident. Which may be exactly what it is, an ecosystem created and maintained by life for its own ends.

To appreciate fully the extent to which life is responsible for the Earth as we know it, we need to imagine what our planet must have been like before life arrived, around 4,000 million years ago. The solar system was just settling down, fizzing with radioactive energy left over from the explosion of a neighboring supernova. Debris from the disaster continued to collide with the coalescing planets around our Sun, splashing molten rock and overheated gas about like the brew in a witch's cauldron. This was the period in our history now known appropriately as the Hadean.

It was a nightmarish time. A time before cohesion. Though the Sun was young and cool, our planet's interior was three times as hot as it is today. Volcanism was rife and made the early oceans warm and rusty. There was an atmosphere dominated by carbon dioxide, so thick and hazy that the sky would have been permanently orange, the sea a shade of dirty brown. The blues and whites we have come to associate with our wispy planet are a more recent product of the bleaching power of oxygen, but there was none loose then. Only a greenhouse of fetid gases oozing sugars and amino acids. Just right, as it happens, for life to begin.

It started simply. Solid evidence for that time is scarce, but it seems clear that there were sufficient organic chemicals lying around to produce and to feed the first microorganisms, which were probably bacteria. In all likelihood a variety of different kinds of early bacteria, taking advantage of local abundance, trying out new strategies, feeling their way into the cracks and crevices of the planet's emerging ecology. These microbes became, and still are, the "silent majority" of life on Earth. There are more of them than all the rest of us put together. They run all the basic industries. It is bacteria that release nitrogen from the soil and make hydrogen by encouraging fermentation. Methane and ammonia are the products largely of bacterial activity, as are over forty other minor gases which together make up Earth's anomalous atmosphere. All of these are "biogenic," produced entirely by living things, most of them microscopic, that churn out these vapors without pause and in astonishing quantities. These were the first agents of our self-replicating system, the ones largely responsible for colonizing the Earth. And as soon as one of them discovered how to feed, not on the limited raw materials around them, but on the inexhaustible supply of energy available in sunlight, our world turned from being merely a dead planet carrying life as a passive passenger, to something that gives every appearance of being altogether and very much alive.

Work on the way in which life directly influences the environment is going on apace, and some of the results are surprising . . .

Reef-building corals, for instance, need warm sea water, but not too much or too little of it. They cannot survive exposure to the air and are easily destroyed by wave action, so they do their best to produce areas of calm water at a uniform depth. Many species release a fatty substance that hydrolyzes, forming powerful surfactants. These, despite the name, depress rather than encourage surf buildup. They alter the surface tension of the sea, pouring oil on troubled waters, reducing wave height and wind stress, making the ocean around fringing reefs far calmer than it would otherwise be.

This effect is obvious to anyone sailing in the tropics. Even at a distance, I can tell the difference between coral and rock submerged at the same depth. The sea over a living reef is far calmer than that which washes around a lifeless outcrop. And remote sensing from orbiting satellites shows that the same principle works on a grander scale, producing unusually short wave lengths in coastal areas where coral abounds.[103]

In the open ocean, something similar in the way of environmental control is achieved by phytoplankton. These tiny green algae float at the surface of the sea, feeding on the sun and doing what they can to ensure that their world doesn't get too warm. One of the byproducts of their metabolism escapes into the atmosphere, producing submicroscopic aerosols of sulfate salts which serve as nuclei around which water vapor can condense. Without these seeds, there would be few clouds that far from land. With them, clouds form readily, solar radiation is reduced, sea surface temperature is controlled, wind blows more strongly, the ocean is more often disturbed and nutrients from the depths are brought back to feed depleted surface layers[67].

On the land, moist tropical forests and their microflora and fauna exercise similar constraints. Seventy-five percent of all rainfall is recycled from leafy canopies and damp floors, encouraging evaporation, creating local clouds which shield the area from excessive radiation. Each of an estimated 30 million species living on Earth plays its own part in such processes. The Amazon's vast army of ants, as an example, produces an estimated 200,000 tons of formic acid a year, spraying this into the air as part of their communication system and defensive behavior. This makes rain in the river basin mildly acidic and promotes the decomposition and recycling of dead wood, releasing bacteria into the atmosphere, where they in turn act as ice-nucleating agents, extracting more rain and more acid from passing clouds. And so on . . . [338]

We are dealing, I believe, with something more than a succession of happy accidents. The origin of life may have been serendipitous, a physical and chemical coincidence that, according to the laws of chance, was bound to happen somewhere, sometime. It may have happened many times before, but in our case on this planet, the experiment has been wonderfully successful. It continues being so because a lot of feedback loops have developed here to prevent life from disappearing, as it may well have elsewhere, making our world more hospitable to it.

Our world is finely wrought and delicately balanced. It is filled with surprising coincidences and the sort of unnatural instability that plucks order out of chaos. It has an extraordinary persistence, a lovely symmetry, and so much numerical coherence that it becomes hard to accept its existence as just an accident. The most astonishing discovery has been that much of it is far from the equilibrium which entropy requires and that it succeeds, somehow, in staying that way. The tuning involved in this is so precise that even our entire galaxy, which was once thought of as merely an inanimate collection of matter, now takes on many of the aspects of life.

Individual stars have a finite life. When they have burned all the hydrogen in their cores, the outer layers expand to turn them into massive "red giants" 100 million miles in diameter. This is due to happen to our Sun about 5 billion years from now as it joins the general tendency to slide towards the stability which is reached when everything has been turned into end-of-the-line, lumpy, inert Iron 56—the substance with the lowest known nuclear energy.

Things go that way because Iron 56 is the most efficient way to pack and store a collection of matter. But if the star involved in this preparation for shutdown is large, more than eight times the size of our Sun, there is one further step that can be taken. If the weight of Iron 56 is great enough, the star succumbs to gravity. Electrons and protons are forced to blend into neutrons, the packing gets tighter, the bottom drops out and star stuff is accelerated so fast that it blows itself apart.

All hell breaks loose in the form of a supernova. An exploding star that recycles itself, sending its matter rushing out into space where it provides the raw material for making new stars and new star systems. But it can only do this because it too depends on the Goldilocks Effect, on the nature of one of the fundamental forces in the universe being "just right."

If the weak force were any weaker than it is, matter in the form of neutrinos would just slip quietly away into space. And if the weak force were any stronger, neutrinos wouldn't be able to go anywhere at all. As it is, they rush about in precisely the right way to create the sort of explosion that brings clouds of bright new star stuff into being. This is necessary for galaxies to grow and evolve. It is the cosmic equivalent of sexual reproduction which, in organic evolution, produces variety by redistributing genetic material. Supernovas seed the galaxy, turning it into a star nursery where a variety of suns grow, many of them maturing into supernovas themselves, carrying on the lively and creative lineage. But there is also a dark side to it all.

If a dying star is too large, gravity overcomes everything. Above a certain critical mass, spacetime becomes bent, escape velocity exceeds the speed of light, and the star simply shuts itself off from the rest of the universe. It becomes a "black hole," one of several hundred million that already punctuate our galaxy, sucking up loose gas and dust like vacuum cleaners whose bags never need changing.

These "tunnels to nowhere" have an inevitably sinister aspect, threatening those who get too close with nonexistence, but it is important to understand that they are inevitable, perhaps even necessary. The same fine-tuning that makes our universe "just right" for life also encourages the production of black holes. John Gribbin, noting that the universe is "so efficient at the job of making stars and turning them into black holes that it could almost have been designed for the job," was the first to suggest that the whole thing might be a black hole itself.[161]

Viewed in this light, if one can talk that way about something that swallows light, black holes graduate from one-way tickets to oblivion to being seeds of new universes. The result of one of an older generation of black holes going about its natural business of reproducing itself. Which suggests that our universe, in its turn, may have been born in just this way, out of a black hole somewhere else. And if the analogy with sexual reproduction is the right one for this process, it is possible that each time a new universe is born it alters the rules slightly, mutating in the way that life does, setting up the possibility of competition between a whole generation of related universes, which opens up the way for natural selection to work amongst them, favoring those most likely to survive and to reproduce again.

As a biologist, this is an idea that appeals to me. I like the notion of laws of nature being subject to selection and evolution in the same way as the species they produce. There is a nice symmetry to that, and it does help to explain how and why our universe should be so unlikely. Its strangeness becomes inevitable. In Gribbin's words: "The fact that our Universe is 'just right' for organic life forms like ourselves turns out to be no more than a side effect of the fact that it is 'just right' for the production of black holes and baby universes."[161]

And if we accept the idea of a living universe, one involved in the struggle for survival with others of its kind, we must allow our universe to be part of a larger ecology in which all are subject to natural checks and balances. To laws and conditions which, if they work in the way that ours do, will produce creative tensions by setting up rival forces that build up and break down, that encourage equilibrium and succumb to disintegration, finding, if they are fortunate, their own proper solutions. Their own way of taking advantage of the Goldilocks Effect and getting things "just right."

In our situation, I have dealt so far only with life-enhancing coincidences, but there is another rival set of influences that subject the system to sudden, unpredictable stresses in the form of catastrophes that can be, and probably have been, the result of mass extinctions.

Geologists divide Earth's history during the last 600 million years into eras whose boundaries are marked by the sudden disappearance of large numbers of species. The death of the dinosaurs has been attributed to every imaginable cause, but evidence is strong for a cometary collision with us that threw up enough dust to shroud the planet and thoroughly disturb our gentle equanimity.[310] The discovery of high levels of iridium, an element so rare on Earth that it is usually assumed to be of extraterrestrial origin, at many of the sites and times of great extinctions, suggests that dinosaur doomsday was not unique. It has happened often, perhaps every 26 to 30 million years.[312] We have known for some time that billions of comets circle our Sun beyond the orbit of Pluto in a frenzy called the Oort Cloud. Now it seems that something disturbs this swarm on a regular basis, sending parts of it hurtling into the space we occupy, penetrating our atmosphere and stinging Earth's surface.[100]

What gives rise to this circle of unrest is still open to question. Some astronomers favor a "dark star," an unrecognized companion to our Sun, which has an eccentric orbit one or two light years away that comes close enough to this solar system to upset it from time to time.[414] But no such celestial stalker has yet been spotted. Others suggest that the cycle of extinction may be more fundamental, something galactic, connected perhaps with a regular oscillation of our solar system in and out of the plane of the Milky Way.[311]

I do not think there is any need for this force to be an actual companion star, but I do think that it is inevitable. Without something like it, the system doesn't work. With it, we have a new mechanism for a problem that has long confounded biology.

Ever since Darwin, evolution has been seen as a struggle for survival, as a long slow race in which all life's competitors are required, like the Red Queen in *Alice Through the Looking Glass,* to keep running just to stay in the same place. The metaphor is familiar and persuasive, but it has never seemed appropriate to me or for the nature I know. The picture of creatures striving to improve themselves, climbing steadily up through the net of evolution, competing with each other but careful always not to get too far ahead of their peers, is far too respectful, too Victorian for my taste. There may well be such an internal dynamic, but there seems always to have been something missing in the traditional Darwinian description, something a little less predictable, more unruly. Something that recognizes life's capacity to surprise and to cope with the surprising.

The dark force, whatever it may be, represents the unconventional, the capricious, and suggests that advance occurs more often as a result of profound change. The sort of thing perhaps which happens when catastrophe strikes, when normal rules of engagement are suspended and some new code has to be introduced to take their place. I see evolution, which reaches always for the Goldilocks solution, seeking rightness in the middle ground, dealing very well with such disaster. Species may disappear, but life goes on with new and more resilient forms stepping up to take their empty place.

This may not be easy. Evolutionary omelettes don't get made without breaking some taxonomic eggs, but it seems that life as a whole is very hard indeed to kill. That may be bad news for environmental organizations, perhaps even for the human species, but it seems closer to the truth than some of the more alarmist scenarios which bring our world to a quick and untimely end.

Struggle is not new in the universe. Stars are battlegrounds for the opposing forces of gravity, which tries to crush them; and thermal pressure, which seems intent on blowing them to smithereens. The violence involved is awesome, but not necessarily destructive. Our universe was born in an outburst of unlimited ferocity, which paved the way for creativity. The forces involved were real and morally neutral, representing neither good nor evil, but it was always important that they should be more or less evenly matched. Without that element of equilibrium, all activity would long since have ceased. And I believe that, even at our own small, personal level, such balance remains essential.

My intuition is that "evil," for all its dark and threatening aspects, is inevitable. A sort of black hole in nature. Get too close and you get sucked into another reality in which all the rules have been changed. The form and character of particular evils may be determined in this reality, which in turn inherits its pattern from another. And "good," I suggest, is not necessarily the opposite of evil, but one part of the field in which both exist. That part which straddles a fine line along which things are "just right." Being there encourages optimal growth and perfect proportion, providing neither too much nor too little, but "just enough" of what we need to keep out of trouble and survive.

I am aware of making a value judgment in assuming that good is better than evil. I am aware of how evil profits, at least in the short term, and can imagine an alternative reality, one of those universes on the other side of a black hole, in which evil holds permanent sway. But the image gives me no pleasure and I feel convinced, from what I know of natural history alone, that such a universe will not be selected, even under a different set of rules. It lacks biological cohesion. It fails to satisfy or to show any of the creative qualities which contribute to cooperation and the sort of coevolution that has produced something like our world. It is, in a word, "ugly." And my guess is that this is more than a subjective assessment. I think we are good enough products of our own evolution to be sensitive to that which does not quite fit, which falls short of the sort of inclusive "fitness" which is necessary for survival.

So in the argument that follows, I intend to pursue the biological consequences of the Goldilocks Effect and of the idea of "rightness" as a force in organic evolution. But I also want to see where "wrongness" leads in life and why, if it does not have survival value, there is still so much of it about. There have to be valid natural reasons for the continued existence of evil. Reasons persuasive enough to help us understand why the devil should still have so many of the good tunes . . .

The answer, I suspect, lies in taking a closer look at the way in which life interacts with its whole environment. A look at ecology, on a global scale. Everyone talks about ecology, but few take the trouble to find out exactly what it means. Which is a pity, because of all the sciences it is perhaps the most immediate and accessible, touching every aspect of our daily lives.

Ecology is concerned with two basic problems: distribution and abundance. Where things live, and how many of them manage to do so. Ecologists set about answering these questions by studying two natural units: individuals and populations. A population being all those individuals of a species that live in a particular area. And because no organism lives in isolation, the science of ecology builds up a picture of how individuals and populations interact, how they exist in a community, and how such communities combine with climate and geography to form a stable ecosystem.[231]

In short, ecology looks at *distribution, abundance* and *association* and tries to discover, through these factors, how the world works.

This is the part that intrigues me. And to relate my look at the evolution of evil to the techniques and discoveries of ecology, I need to make just one basic assumption. It is perhaps a value judgment, but we have to start somewhere with some sort of baseline. So I am assuming that a sound ecology, a system that is reasonably stable and works well in the common interest of all of its component species, is something to be desired. A "good" thing. And that any factor, actor or influence which disturbs such an ecology is worth avoiding in the short run. It is a source of confusion, something not to be desired. A "bad" thing. There are instances in which such influence may turn out, in the long run, to be beneficent. To produce changes that alter the course of evolution in some even more productive direction. We will deal with these later, but for the moment I am asking you to accept the idea that some things are "right" for life and some things are "wrong." And I ask you to follow me through a sometimes tenuous chain of ecological examples which will, I hope, demonstrate that it is possible to identify the good from the bad, the right from the wrong, and discover how evil came into being.

My doctoral days were spent very happily and productively under the guidance of Desmond Morris at London Zoo, which in the 1950s was still the best and most representative collection of animals in the world. Whenever a question arose about behavior and its difference in closely related species, I could go out into the gardens and compare the way in which gelada and chacma baboons greeted one another, or scarlet and hyacinth macaws dealt with the same rock-hard palm seed.

Such easy access was an enormous luxury, but it also created certain problems of perspective. I soon found myself thinking and talking, not about where a particular species lives, but where it "comes from." This is a problem familiar to zoo people and I had to keep reminding myself that the collection in which I worked was unnatural, little more than a popular monument to the role of dispersal in distributing life around the planet. As soon as my thesis was complete, I escaped from the gardens and went to find giraffes at home in Africa, wombats in Australia and polar bears in the Arctic, where I discovered that geographic, climatic and behavioral barriers keep them confined almost as effectively as the bars in a zoo.

There is no biological reason why polar bears should not thrive in the Antarctic, merely a good earthly reason for their inability to get there without help.

Disasters play a role in rearranging life on Earth, but in the short run it is more often we humans who upset ecosystems by removing crucial predators or adding problematic pests. Some of these have disturbed the equilibrium so badly that they deserve to be considered as wrong, perhaps even as potentially evil.

On Christmas Day in 1859, for instance, a sailing brig with the ominous name of HMS *Lightning* arrived at Melbourne in Australia with a dozen wild European rabbits bound for an estate in western Victoria. In 1862, a bush fire destroyed the fence around their enclosure and within thirty years rabbits covered 1.5 million square miles of the continent, living in environments ranging from stony desert to alpine valleys, from coastal forest to the grasslands of the interior. Billions of rabbits have since been shot, poisoned and infected with lethal viruses, but even such drastic measures have done little to control what amounts to one of the most disruptive invasions ever known.[282]

The introduction of rabbits to Australia was "good" for this one species, *Oryctolagus cuniculus,* which now ranks as one of the most successful mammals on Earth. But bringing it to that island continent, where none of the usual checks and balances exists, has been like flooding a peaceful rural community with well-trained city pickpockets. It has been "bad" for Australia, an ill-considered act, ecologically and morally unsound.

I base this assessment on a set of common suppositions which ecologists make, but seldom put into words. First of these is that all plants and animals have a place of origin, a *locus.* This is where they "belong," where they fit. Such fitness is generally seen as an ability to become part of an ecological web. They coevolve with other species in that place and assume a position of least resistance, not consuming too much nor laying themselves open to undue consumption. They become "just right."

When a species wanders beyond its original bounds, it most

often does so slowly. It migrates or is displaced by some natural event such as a flood or a storm, swimming, flying or rafting to a new place, "putting down roots" there, competing with native species for available resources. Most such invasions fail because local species are better adapted to their locus, but sometimes invaders have an unfair advantage. They break the rules of coexistence and get away with it because they have gone beyond their usual jurisdiction. They are no longer subject to the fine constraints of home. They step out of line and find that there are no penalties for doing so. Nobody eats them.

So Asian tamarisk sucks American semideserts dry, draining down delicate water tables, eliminating ponds and all those native species which depend upon them, depositing new kinds of chemicals on the surface of the soil, changing the very shape of the land by encouraging erosion. African bees used to tougher circumstances outmaneuver local species, chasing them out of their homes, interbreeding, colonizing South and Central America, becoming "killers" in the absence of their usual constabulary, the African bee-eaters. Eurasian zebra mussels, free of their usual predators, clog waterways around America's Great Lakes. And in California, the Mediterranean fruitfly even succeeds in driving a governor from office.[337]

A slow-motion explosion is taking place as natural barriers fall, most often with human help. We have an unfortunate habit of taking more than our personal baggage with us when we move. There was even a time, before bubble-wrap, when we packed our possessions in straw or other vegetable padding, sowing dissension as we settled, giving a head start to dozens of carpetbaggers. Old World plantains which arrived in the New World with the Pilgrim Fathers followed them so faithfully that they became known to native Americans as "White Man's Footstep." Sometimes such hitchhikers fail to find a foothold in their new homes, but more often we give them an additional helping hand by degrading the local system—felling, clearing and cultivating until the invaders triumph. Such biological or ecological "imperialism" is serious and has effectively destroyed the native vegetation of much of Hawaii and most of the Marquesas Islands, and brought over 1,000 alien species into New Zealand.[206]

These are unnatural acts, outlawed now by most nations because local governments have discovered that such invasions are expensive. They cost money to put right—an estimated $134 billion in the United States alone. But there are other, more biological reasons for considering them immoral. One of which is that they needlessly disturb the delicate equilibrium of Earth. They may be right for Scotch broom and Australian pine, which get to see the world, but they are wrong for the planet as a whole. They are unfair.

I want, with the use of notions such as "rightness" and "fairness," to widen concern from individuals and populations to the ecosystem as a whole. The Goldilocks Effect, nature's way of getting things "just right," works well for feral rabbits and little girls sponging on gullible bears, but it only finds its proper level, gaining evolutionary significance and satisfying Aristotle's criteria for proper behavior, if we look at it in action on a larger canvas. Only then can we see how and why it succeeds or fails, but even then the mechanism is far from obvious.

Rabbits reached Australia with the first European settlers as early as 1788, and repeated introductions took place with every new wave of colonists. By the beginning of the 1800s rabbits were common in every village and were liberated, both accidentally and deliberately, many times. All without alarming consequences, until the 1859 introduction that has proved, for reasons still difficult to identify, so disastrous. Evil, it seems, may have its own peculiar reasons and seasons. And it would be very useful to know exactly what these are.

I believe that the best way to uncover the biological roots of being right and being wrong may be to introduce a new area of ecological and ethical enquiry. This deals with evil in action and takes its name from the Greek root *pathos,* meaning "suffering." Pathos is the opposite of ethos, which deals with nature, character and community, giving rise to "ethics" and "ethology." Pathos instead produces "pathology" and has come to imply sadness and to excite pity, but the adjectival and substantive form "pathic" has never before been applied, except as a rare and now obsolete description of those who submit passively to sexual abuse.

I choose to revive and use *pathic* and *pathics* here in a more general, direct and active sense, to identify the study of that in nature which represents a loss of character and community, leading away from rather than towards natural cohesion.

Living systems are not static, they move and grow and in doing so become vulnerable to distortion and deflection by both internal and external factors. I want to be sure that we understand how such biological influences might work on the ecosystem as a whole and, to that end, am proposing a set of rules that may help. There seem to me to be three major areas of action involved and I am calling my formulation of these the Three Principles of Pathics. These are, in effect, a description of "How Good Things Get Bad." And if ethics deals with rules of conduct, then pathics is concerned with misconduct and ought to have rules of its own. These are not just the opposite of "good," but involve a series of subtractions from order and stability, and the first of these may be phrased very simply in this way: ORDER IS DISTURBED BY LOSS OF PLACE.

In other words, stability suffers when something is removed from, detached from or distanced from, the locus where it works best and set down somewhere else, where it is no longer part of a larger system of mutual advantages and constraints.

At a superficial level, this is a rule so self-evidently right that even bureaucrats recognize its value and prohibit the introduction of exotic plants and animals from one country or continent to another. But I expect more of this First Pathic Principle than customs control.

I see little biological difference between the rabbit colonization of Australia and the invasion of our own bodies by microorganisms given the chance to establish themselves in such a promising new environment. Pathology, the study of disease or "dis-ease" in any organism, involves morbid conditions that arise due to an essential imbalance, often produced, or taken advantage of, by an invading pathogen. We have learned to deal with such attempts at colonization much as plants do, by chemical warfare, often using the same botanical remedies, borrowed for application in the form of antibiotics. But sometimes the best defense involves another living organism, one better designed to compete with the invader.

Hot on the heels of the rabbit plague in Australia came another disastrous introduction. This time it was a plant, the American prickly pear cactus *Opuntia,* imported by early settlers who valued them as ornamental shrubs. In their native habitat they also provide useful and easily controlled fences for enclosing livestock, but in Australia they ran rampant. By 1925 more than 100,000 square miles of Queensland and Victoria were so thickly covered by cacti that the land was useless. The cost of removing the prickly invaders mechanically or chemically was more than the land was worth, so entomologists looked back to America for a natural predator, an ally that might help do the job for them.

They found it in the form of *Cactoblastis,* a little moth whose name gives a sense of its aim. It has a voracious appetite for *Opuntia* and without the American birds which traditionally keep them in check, hordes of its caterpillars marched across the spiny wastes of Australia and cactus melted away in their path. Just five years after their introduction, the prickly pear problem was under control and has stayed that way ever since.[115]

This episode is a timely reminder of how large an effect even a small herbivore can have on a major plant population. It also illustrates an important feature of the First Principle of Pathics, which is that it always starts from scratch, working on what is available. It has no memory and no long-term plan. It plays no favorites. Displaced cactus disturbed the nineteenth-century order of part of the Australian ecology. But by the twentieth century, prickly pears were the dominant species in that ecology and had become the new order that had to be disturbed, in its turn, by a further displacement. What was bad for *Opuntia* was good for the larger ecology of Queensland. There is no room for sentiment in either pathics or pest control.

It is clear that it is not going to be easy to attribute blame or to make ethical judgments in these areas that will be unequivocal. Evil, like much of the rest of life, is relative. And we will be forced very often to resort to the kind of analysis which makes mathematical instead of moral assessments. Which is why I have tried to ensure that the principles of pathics are truly neutral and do no more than describe a process—in this case one which shows how good things can get bad, even so bad that they overwhelm us. The degree to which we now disturb Earth's ecosystem is so great that we very badly need to find ways of measuring and curtailing what we do; but the right way, the Goldilocks solution, is not that easy to identify.

Giant sequoia, the spectacular redwoods we so much admire, the ones that seem to dominate old-growth forest in coastal California, are in fact a product of disturbance. They require bare, litter-free soil to germinate successfully and thrive on the sort of frequent siltation and flooding that kills competing grand and Douglas firs on river flats. As mature trees, they have thick, flame-resistant bark and are seldom damaged by ground fires that are fatal to salt and sugar pines. So by protecting forests from fire and flood in national parks, we have signed the giant trees' death warrants, condemning them to be overwhelmed eventually by less resilient, but more numerous, competing conifers. We are killing the giants with kindness, succumbing to the sentimental entreaties of that misguided arch-propagandist Smokey the Bear. When what we really need is "just enough" of those staunch old evils—fire and flood. Forest managers in California have now learned to live with them, letting fire and water take their natural course as often as they can.[229]

When in doubt, the best course of action seems to be to leave things as they are. It is alarming how quickly something benign can become malign when things are disturbed. Recent epidemics of yellow fever in areas with no previous history of the disease can be traced directly to reservoirs of the infection in a few monkeys living in the forest canopy. They and their mosquitoes have long reached an accommodation of relative immunity up there, but felling the trees has brought this aerial ecology crashing down, breaking the reservoir, exposing human loggers and new insect carriers at ground level to the pathogen—with disastrous results.

This is, of course, another example of the First Principle of Pathics at work. Order has been disturbed, creating an ecological problem by displacing mosquitoes and their parasites from a system in equilibrium to one in which there has been no time to develop appropriate controls. Good has been made bad by moving it somewhere else. What was "right," or at least in balance, in the canopy, becomes "wrong" in a terrestrial community unprepared for change. In some areas the new hosts died along with the displaced parasites. But in others, an extra factor has entered the equation. Loggers in Brazil, Malaysia and New Guinea, where the new outbreaks of fever have taken place, live in sprawling frontier communities with large numbers of people in close proximity to each other. Ideal feeding and breeding grounds for mosquitoes, and for the transmission and spread of the infections they carry. Places ripe for epidemic disease in the same way as the large settlements of the early neolithic period, and the new cities of the great empires, were for the first time vulnerable to preexisting organisms which now took on the form of plagues.

Which brings me to the Second Principle of Pathics. Where the first principle was qualitative and addressed ecological concerns about distribution, introducing the idea of right and wrong *places,* the second is largely quantitative and concerned with the matter of right and wrong *numbers.* This principle deals with the ecological problem of abundance and what it means to have too much, or too little, of anything. It has to do with resources and with population and may be phrased in this way: ORDER IS DISRUPTED BY LOSS OF BALANCE.

Everything is limited. And in biology, this simple observation is nowhere more obvious or more vital than where it touches population numbers. It is one of the central understandings of ecology that no population of any species can increase without reaching limits imposed by its environment or its own behavior. And very often the mechanism involved in establishing such control is astonishingly complex. Nowhere more so, perhaps, than in a rodent Norwegians call the *lemming.*

Lemmus lemmus is a small, thickset mouselike animal with a blunt, rounded muzzle, small ears, short tail and a coat so thick it seems to have no legs at all. It lives on the edge of Arctic tundra, a habitat covered in snow for much of the year, but does not hibernate, tunneling instead beneath the surface where it is protected from extreme cold, feeding on roots and the frozen buds of mosses. In all this it differs very little from eight other species that inhabit the far north of America and Eurasia, but there is one aspect of the Norway lemming's behavior which has given this little mammal almost legendary status.

Every few years, regimented masses of lemmings descend from mountain heaths and invade forests and pastures where they "destroy crops, foul wells and, with their decomposing bodies, infect the air, causing terror-stricken peasants to suffer from giddiness and jaundice. Driven by an irresistible compulsion, not pausing at obstacles—neither man nor beast, river nor ravine—they press on with their suicidal march to the sea."[199]

These population explosions are real and so massive and sudden that Scandinavian folklore insists that lemmings are not of this world, but generate spontaneously from foul matter in the clouds. One of the Inuit names for them means "creatures from outer space," and a sixteenth-century Swedish woodcut shows lemmings falling with heavy rain directly from a threatening storm.[251] They don't, of course, and neither do their cyclic migrations end inevitably with mass suicide in the sea. They certainly do move in very large numbers, at random to begin with, but where their progress is blocked by a lake or funneled by mountain walls which concentrate the flow, the accumulation becomes so great that panic ensues. Then the lemmings take to extraordinary and reckless flight, across rivers and glaciers, over cliffs and occasionally into the ocean, where they do sometimes drown in very large numbers indeed.

The reason for this periodic urge to behave mindlessly, "like lemmings," remains largely mysterious, but the progress to hysteria is now well known. Each cycle begins with lemmings at a low density, perhaps just one animal to every five Arctic acres. But as the snows melt and animals come into more frequent contact, breeding begins. Females produce a litter of up to eight offspring every three or four weeks, and the young themselves begin to breed when they are just one month old. Soon the population density rises swiftly to ten or twenty per acre and snowy owls, foxes, weasels and even wolves begin to show an interest. In most years this is enough to keep the lemmings in check, but the "balance of nature" is nowhere as simple as we once believed. Sometimes predation on its own cannot cope and other restraints become necessary when, every three or four years, lemming numbers soar to well over a hundred an acre. When that happens, it seems, the lemmings themselves turn mean.[261]

Even at the best of times, Norway lemmings are an intolerant lot, given to brief encounters in otherwise largely solitary lives. But as density increases, so does aggression. The change appears to have nothing to do with competition for food, which remains plentiful in the short but abundant Arctic summers. It has everything to do with proximity. As lemmings are forced into more frequent contact, they fight. Males box and bite each other, females are ferocious in defense of their young. Wounds become common and can prove fatal. Adults of both sexes take to killing small lemmings soon after they are weaned. Young animals fail to reach sexual maturity until they are older and larger than usual, and males that do succeed in reaching puberty are immediately attacked by their older counterparts. By midsummer, many females stop breeding altogether because it no longer pays them to raise litters that are going to die, and this slows the population growth but does nothing to halt the rampant aggression. Eventually, the pressure gets so great that young animals who have been repeatedly defeated by older ones start to move out and one of those spectacular lemming migrations begins.[339]

The important feature of this cycle is that normal order is disrupted by loss of balance in the population. Something has to be done to limit population increase and in lemmings the limiting factor is aggression. Aggressive individuals succeed in crowded conditions and selection favors those who continue to show the greatest aggressiveness. In this species, population levels depend to a certain extent on food, predation, climate and disease, but the ultimate sanction is aggression associated with overcrowding. It is good for lemmings to breed successfully, but bad if they do so too well. The Second Principle of Pathics draws attention to the disruptive effect of such imbalance and it is significant that in lemmings at least, the cure for disorder happens to be a dramatic increase in behavior such as assault, battery, child abuse, murder, infanticide and finally suicide.

Much the same may be said for crowding in cats, which leads to an almost continuous antisocial frenzy of hissing, growling and scratching. And for crowded rats, who descend even to cannibalism. All behaviors we legislate against in our own species and tend to ascribe to the evils of urban life. But it would be wrong to assume that smaller populations are always better for a species.

In the early nineteenth century there were hordes of pinkish birds, like turtle doves with long tails, in the woods of North America. They wintered in the southeast, in Kentucky and the Carolinas in flocks which the naturalist John James Audubon estimated as a billion strong: "obscuring the light of noonday as by an eclipse."[17] And they bred around the Great Lakes in nesting groups covering a hundred square miles of forest at a time, loading every tree so heavily that the boughs often broke. These were of course the late passenger pigeons, destined to die in their billions, clubbed to death relentlessly by night in their roosts, suffocated by the gas and smoke of sulphur torches or shot for pig food. Ornithologist Alexander Wilson saw 2 billion birds pass overhead one day in 1808, but a century later there was not a single passenger pigeon left in the wild and the last known individual died on September 1, 1914, a captive in Cincinnati Zoo.[143]

It is almost inconceivable that the largest social assemblies in natural history could have been hunted into extinction so thoughtlessly, or so easily. Most species are more resilient, but there was something about the passenger pigeon that made it particularly vulnerable. It was intensely gregarious. Its entire social structure depended on large numbers and on close proximity, and when its population fell below a critical threshold, whose value we will never know, its viability was undermined and destroyed. The last passenger pigeons died, not so much by fault as by default. You *can* have too little of a good thing and the Second Principle of Pathics must be seen as predicting the disruption of order by a loss of balance in either direction. Abundance is not a description of plenty, it is a measure of quantity equally applicable to "a lot" or "a little." Underpopulation is as bad as overpopulation, too many resources as harmful as too few, and stinginess as great a sin as greed.

Aristotle was right. Bad desire consists of wanting, not just too much, but also too little. And anything which isn't "just enough" or just the right number for an individual, a population, a species, a community or an ecosystem is bad for the whole and evidently evil. And if limits, both high and low, apply to all living populations, then humans cannot be exempt.

Through most of human history, we have been a "rare" species.[74] Our populations rose and fell with the usual seasons, growing when times were good and slipping back when climate, disease or self-inflicted wounds of war took their toll. The shift was seldom more than 0.01 per cent up or down in any given year. We were as stable then as chimpanzees or checkerspot butterflies. Egypt, for all its turbulent history, had the same number of people scattered along the Nile in 2500 B.C. as it did in 1000 A.D. But in the last millennium, everything has changed.[114]

We differ from most other organisms in being able to exercise our own controls rather than relying on natural mechanisms. Advances in health care, housing and nutrition have all contributed to a change in lifestyle for enough of us to produce a population growth spiral for the whole species, with increases of up to four or even five percent per annum. But unfortunately these go largely unchecked. We have reduced the death rate without an equal reduction in the birth rate, and the sheer weight of our numbers is threatening not just our survival, but the order and ecology of the entire planet.

And it is not just we who are disrupting the balance, but our creations—those genetically engineered organisms we have been introducing into natural communities for centuries. The worst of these is perhaps the domestic goat, altered by selective breeding from its wild ancestors to a ruthless eating-machine that became a major factor in the destruction of its own ancient habitat in the Mediterranean. Not content with that, we took it with us on all our early exercises in empire-building, leaving it behind to breed, and to produce useful supplies of meat and milk, on remote oceanic islands. I have seen the destructive result of this displacement and disruption, this blatant promotion of the First and Second Principles of Pathics, on Guadeloupe and St. Helena, on

Timor and Antigua; and have come to recognize, even from a distance at sea, the sort of barren wastelands, the devastation of once wooded and well-watered communities that ecologist Paul Ehrlich so aptly describes as "goatscapes."

It is probably no accident that the devil is so often and so widely seen as shaggy and horned, taking on the form of a black goat; portrayed with cloven caprine feet as Pan or Dionysos; or identified as the apocalyptic goat-beast of Christianity and the hairy-bearded *se'irim* of Hebrew iconography.[330]

We have, in furthering our own good, far too often become instruments of evil, disturbing the ecology by clearing forest, draining marshes, polluting air and water; but it would be a mistake in all this to assume that nature has a perfect form and exists, in our absence, in sleepy equilibrium. The famous "balance" of nature is an illusion, one that disappears very quickly when you look too closely at it. Fluctuations in natural populations are commonplace, the rule rather than the exception. Birds and butterflies that are abundant one year may be gone in the next. Schools of little silver fish—herrings, alewives, shad or menhaden—which mark the seasons by their shimmering migrations, sometimes fail to show up at all. And we have had to lose our nineteenth-century certainty in natural constancy. There may well be evolutionary checks and balances in a system, but to identify them we are going to have to uncover far more elaborate and interconnected, far more interesting, models of control.

Among the first of these were a set of equations developed in the 1930s by Vito Volterra, who had a strong influence on the development of calculus. Volterra was a professor of mathematical physics at Rome and a senator of the kingdom of Italy, but with the rise of Fascism and the end of the monarchy, he traveled abroad and turned his interest to living systems, trying to establish patterns which allowed tyrants and people, wolves and sheep, to coexist. What he produced was an abstract predator-prey model that was fascinating and counterintuitive.

The Volterran principle predicts, for instance, that if the populations of predators and their prey are more or less in balance with one another, and something happens to disturb their joint environment, then there will be a disproportionate decline in the number of predators. Surprisingly, this is precisely what happens. If herbivorous insect pests in a crop field are being eaten by other predatory insects and the farmer sprays the entire area with a broad-spectrum pesticide, a similar proportion of each population will die. But the crop-eating prey, whose death rate has been reduced by a shortage of predators, recovers more quickly than the predators, who can no longer find sufficient prey. So the farmer's even-handed assault on the system seems only to make his pest problem worse. This is not what he intended or what we might expect, but it happens all too often, both with pesticides and to our ideas of how we think things ought to work.

For almost two centuries the Hudson Bay Company in Canada has been keeping records of the furs sold to them by hunters and trappers who work on islands in the bay. These purchases consist largely of lynx and their favorite food, the snowshoe hare. The records show that populations of the two species fluctuate, with a peak of abundance and therefore of fur sales, about every ten years. The population curves more or less coincide, with lynx numbers rising to their high point slightly after the hares. All very nice and tidy, just what one would expect if well-fed lynx tend to raise more litters, who eat more hares, leading to an eventual parallel decline in both populations. But the problem is that on some islands where there are no lynx, hare populations fluctuate anyway and in precisely the same way.

Recent studies have produced a more surprising Volterran suggestion. The snowshoe hare, it seems, is another species that reacts badly to overcrowding. It goes into a form of shock that the researchers diagnose as "hormone-mediated idiopathic hypoglycemia." What that means is that hares suffer from stress which damages their livers in ways from which they cannot recover, even if they are removed from the situation. They have an endocrinal response which controls their own population size. And, as it happens, those snowshoe hares which are most resistant to such stress, and therefore form a larger proportion of the surviving hare pool, also carry a virus that may produce a debilitating lynx disease. This predator-prey relationship is nothing like we imagined. It is actually the hares that assault the lynx.[158]

In any biological community there are shifting configurations, many of which are going to be as complex and as unexpected as this. It is seldom as simple as "who eats whom?," or of removing just those species that we find dangerous or inconvenient. Doing so can cause the community as a whole to change, like pulling the keystone out of a soaring arch brings the whole structure tumbling down. There are such species and their identification has become vital to ecology, which needs to be very concerned about extinctions now that we may be losing as much as a species every second from the biosphere.

As a biologist, I find it goes a little against the grain to rank one species above another, to declare one plant or animal more important, less replaceable than the others in its community. It was thinking of this kind which made human interests paramount and put us into our present predicament. But there is good cause for scientific concern about the welfare of some species whose loss can and does reverberate, making a system far poorer than it would otherwise be. And I am happy to swallow my purist pride where kelp and otters, baobabs and elephants are concerned. Even biologists have their favorites.

Kelps are the sequoias of the sea. Submarine giants who have succeeded in colonizing some of the most dangerous and inhospitable regions of the ocean. They are brown algae whose pigment is appropriate to the shadowy depths, keeping them offshore, chained to the rocks on which they are anchored, when other greener relatives migrated to the more intense illumination of the land. In the surge area below low tide, the kelps still dominate, strong and slippery so as to move in the water without damage, long enough to keep their crowns up near the surface, but not so close that they run the risk of being damaged by sun and air. They are the perfect subtidal plants, "just right" for their habitat, forming forests forty feet tall, providing a habitat for hundreds of other species. Swimming through such algal glades is a profound, almost religious experience. The stalks rise like cathedral columns in the gloom and welcome one in with their gentle undulation. But there is, of course, a serpent in this Eden.

It is called *Strongylocentrotus droebachiensis.* Even from its name, you might guess that its intentions were not entirely honorable. It is a spiny sea urchin, nearly spherical in shape, with well-developed jaws and a huge appetite for kelp. Armies of them gather on the rocky floor when they can, marching as a front through the kelp, attacking the base of the great stalks, pulling out the keystone, bringing the whole forest down in ruins, living quite happily off drift algae on the barren ground that remains.

This is happening now off the coasts of Japan, the Kuril Islands and Kamchatka, with disastrous consequences for near-shore fish and fisheries. All for one good reason: the removal of another keystone species. In this case a significant predator, a sea otter best known for its endearing habit of lying amongst the kelp on its back, cracking an occasional clam on its stomach, but which, more importantly, loves to eat sea urchins. By contrast, off the coasts of Alaska and California, where the otter is strictly protected from fur trappers and therefore more numerous, kelp flourishes and provides an essential nursery area and vital habitat for local marine life.[121]

Given half a chance, things hang together nicely in the cold waters of the north Pacific, where competition is muted by the intervention of a "good guy," an unconscious environmentalist in the form of the sea otter. Without this keystone species, the system collapses. With it, the system is rich and capable of holding far more species than one in so-called "equilibrium," holding only as many species as, and for as long as, predation and open competition will allow. Kelp forest communities are not in balance, but held in a state of delicate and progressive order by the otters, who actually have to work harder than those few of their relatives still left untrapped on the far side of the ocean, where kelp is now rare, but urchins abound and are easier to catch than fish.

So why do sea otters bother? Why work harder than you have to? It is easy, watching an otter dozing off the California coast, one arm wrapped nonchalantly around a frond of kelp to keep him from drifting away on the tide, to accept that this is how things "should be." That otters like hanging around where there is kelp and that their control of sea urchins is casual, one of those happy accidents of nature that keep things "just right." But you may have gathered by now that I am suspicious of such coincidences and try to look beyond them for deeper meaning. And in this case, I believe I see the Goldilocks Effect at work, setting up its antidote to what I call the Third Principle of Pathics.

This rule has to do with the ecology of association, with right and wrong *relations,* and I rank it as the most important of the three pathic principles. The first and second dealt with influences which just "disturb" or "disrupt" natural order. The third principle goes further by suggesting that ORDER IS *DESTROYED* BY LOSS OF DIVERSITY.

The first and second principles recognize that natural order is disturbed by loss of place and disrupted by loss of balance. Good things get bad if they are not in the right place in the necessary quantity. Every businessman knows that. These are the ground rules of economics and ecology—proper distribution and supply. If either or both fail, there are problems of control which lead to dysfunction. The third principle is more general in its scope. It predates the first two principles or presumes that their requirements have already been met. The law of association examines the nature of relations and recognizes that order can be not just upset, but totally destroyed, if connections are impoverished. It matters who you know and how you manage your affairs. And more than that, it matters desperately how rich these associations are allowed to be.

Which brings me, at last, to baobabs and elephants.

I was born in Africa and dream still of savannahs with scattered groups of trees near rocky outcrops which provide the sort of long, hazy views that seem to be part of the genetic memory of our species. But the truth is that both we and the savannahs are latecomers to Africa. Neither would exist at all were it not for the preexistence of another keystone species, one which has been plying its trade for some 50 million years, crafting Africa's landscape and shaping its ecology. This architect on a grand scale is, of course, the African elephant.

Elephants are browsers, in every sense of the word. They feed on woody plants, breaking off branches, leaves and fruit with their elongated noses, giving every impression of thinking deeply while they do it. There is much on their minds. They need over 300 pounds of vegetable matter each day and a home range large enough to find it in all seasons of the year, which can mean covering up to 1,000 square miles, often on well-worn paths or "elephant roads" that crisscross such territories. Their impact on the land is considerable, leaving trees and shrubs uprooted or debarked, killing many species faster than they can regenerate. This habit has come to be described as "the elephant problem" by those who still believe that elephants and trees existed in stable equilibrium until human activity interfered with elephant migrations, altering the size and nature of their territories, and compressing their normal populations into parks too small to hold them. Some of this has indeed happened, but there is persuasive new evidence to support another, more creative, view.

It comes from my all-time favorite tree. A grotesque yet loveable old lady of a tree with fluted skin that looks as though it is melting like wax, running down the stem. It is, to be honest, a fat tree, whose diameter may exceed its height and whose bare, rootlike branches give it the absurd appearance of having been planted upside down. It is in fact remarkably long-suffering, surviving massive indignities such as having a bus shelter or even a whole road carved out of its pulpy interior. The flowers are appropriately blowsy and pollinated by bats, the hollow branches home to hornbills and owls, the fruits a source of cream of tartar and much beloved of baboons; but no one pays more regular attention to baobabs than elephants, who eat every last part of them.

Those trees which do survive can live for great ages, perhaps as much as 4,000 years, and a recent survey of baobabs in Zambia has produced some fascinating results. Age distribution curves for normal populations show a smooth decline, with a large majority of young individuals gradually giving way as members die and disappear for all the usual reasons. But in Zambia, and in most of Africa, it seems, the vast majority of living baobabs range from 125 to 250 years old. There are smaller numbers of older trees and very few younger ones—in the study area, none between the ages of twenty-five and fifty years.[61]

The elephant is the most likely architect of this strange design. And what it suggests about elephant populations is that they fluctuate just like the baobab, probably *because* of the baobab and other trees. There is no such thing as natural equilibrium between elephants and forests. Their two systems cycle back and forth, with elephants increasing in density as they thin out the forest; and declining in density as the trees become too sparse, letting forest regenerate once again. If the baobab is to be believed, elephants were relatively scarce in the eighteenth century, and our recent experience of them has been confined to a period of recovery. In other words, there is no "elephant problem."

Arguments about management strategies continue, but it seems clear that one of the major functions of elephants in Africa, their "job" if you like, has been to convert woodland to grassland, forest to more open savannah. This is disadvantageous to the elephants themselves who, like otters in kelp, have to work harder for a living. But what they are doing is to create new possibilities for other species. The cycles we see in this are characteristic to the African ecosystem and have only been modified in part by recent human activity.

Elephants, like humans, have been involved in massive landscaping projects for a long time, keeping Africa's equatorial ecology from going to its moist tropical climax, opening up the forest enough to let us see the trees, and at the same time opening up new habitats and new opportunities. Understanding this makes it seem imprudent to continue with attempts to manage elephant populations artificially by culling. Such actions are only likely to speed up a cycle that has already begun its own natural phase of decline. Meanwhile, great herds of zebra, wildebeest and gazelle, plus all their predators and followers, have moved into the savannah, which has not only become symbolic of all Africa, but given an ape with ideas the chance to become human.

Given such connections, I can't help wondering how much else in organic evolution has been nudged along its present path in the same way. The process of evolution requires, not a level playing field, not stability nor balance, not equilibrium, but the sort of environmental unrest that makes change both necessary and advantageous. It needs something to work with, some purchase to give it leverage on life. It looks for the bits that stand out from the rest, rejoicing in diversity and limitless variation. Anything which subtracts from that variety is bad for business. Every loss of diversity represents a loss of organic vigor and a corresponding reduction in the possibilities of interconnection and cooperation. And the loss is progressive. As diversity fades, so do the chances for change and eventually the system breaks down altogether.

Lose too much and, like the passenger pigeon, you lose the lot. That would be bad for everyone. Manifestly evil. So pathic that it would make the whole distinction between right and wrong completely meaningless, because both values would disappear altogether.

Diversity is the key to success. It is the peculiar strength of our biosphere that its interconnections—the food webs, symbiotic associations, ecological guilds and kinship groups—are so rich and so complex. Nobody knows quite why things need to be this way, or how such diversity began. It remains mysterious, for instance, that there should be so many more species in the tropics than the temperate zones. It is not immediately obvious how dozens of large tree species, all doing essentially the same job in rainforest canopy, serve the system better than just one species expert at dealing with abundant light, warmth and moisture. But no biologist who has worked in the tropics really doubts that the diversity is both necessary and good.[427]

You can feel it as you walk into virgin forest, a busy silence that is so intense and alive that it fills the mind. Darwin, on his first day in Brazilian forest, described it with feelings of "wonder, astonishment and sublime devotion."[96] Being there is like swimming through a living sea, knowing as you go that all the details matter, that energy flows through this system like a bloodstream in which every capillary counts. We will probably never fathom all such depths, nor is it important that we should. What matters is that they work, and that every naturalist worth his binoculars and hand lens can assure you that they do, even if he doesn't quite understand how, or why he knows this to be true.

Curious naturalists, used to working in the field, get a sixth sense about such things. I can tell from hundreds of yards away whether a patch of woodland on a river will produce a variety of birds, a "party" moving through with mutually enhanced enthusiasm, or nothing at all. I know which stones to turn in search of scorpions and which are most likely to yield a sleeping snake. I can decide, before dropping anchor, which section of a reef, which coral outcrop, supports the greatest variety of fish. There is a completeness, a "glow" to such systems in their prime that banishes doubt. There is health in them, and it shows. Just as systems out of kilter, in which something is not quite right, send out warning signs. These factors, as much as anything, inform my judgment of what in biology constitutes right and wrong, good and evil.

I am concerned about completeness, or the lack of it, in nature. And I am anxious to know if we can learn enough about it to make sense of the part we play. I feel a kind of kinship with otters and elephants and share every thinking person's horror at their thoughtless destruction. The loss of each such asset leaves me feeling distinctly uneasy, somehow less prepared. But I believe that we can make a difference in such matters, perhaps even understand what it is that lets ecocide and genocide take place, if we go back to basics.

The lessons seem to me to be these . . .

Life is elaborately interconnected, largely for the common good. The web, however, is so intricate that some of the links are fragile, which means that things happen, often unexpectedly, some of them good, some bad. Sometimes it is hard, in the short term perhaps impossible, to tell the two apart. But in the long term, and in general, it looks as though "good" is what is right for the whole; and "bad" is what is wrong for the whole. "Evil" is far more difficult to define, but could perhaps best be described as that which is consistently or deliberately bad.

There is clearly a sliding scale between "good" and "bad," but there appear to be three principal ways in which benign things most often deteriorate and become malign:

1. Good things get to be bad if they are displaced, taken out of context or removed from their locus.

2. Good things get very bad if there are too few or too many of them.

3. And good things get really rotten if they cannot relate to each other properly and their degree of association is impoverished.

With these principles in hand, I believe that we have a simple biological framework with which to reexplore the old mysteries of right and wrong. If nothing else, they at least establish some of the rules of the game and make it easier for us to understand what happens when the rules are broken.

In some places recently the rules seem to have been abandoned altogether. Anything goes in Bosnia and Rwanda, and everything goes to hell.

In less drastic circumstances, disarray can be useful, even creative. It is part of human nature to be rebellious; that has always been one of our strengths. But as nice as it may be to be a little wicked, we have to understand the consequences. We ought to know how far we can go without destroying, not just the system but ourselves along with it. We must become aware of exactly who wins, who loses and at what cost.

We need, in short, to put a price on evil . . .

TWO

BREAKING THE RULES: THE ARITHMETIC OF EVIL

We carry a reminder of who we are in every body cell.

A sense of identity is basic to all life, even at a cellular level. It is responsible for all immune and allergic reactions, for the way in which white blood cells reject and remember foreign bodies, producing specific antibodies in anticipation of future invasions. And it is the secret of successful cohesion between all the millions of cells which form part of the same body. Our health depends on it. Otherwise brain and muscle, heart and lungs, kidneys and skin would all have their own agendas and tend to pull in opposite directions.

This is what happens when cancer strikes and some cells go out of control, reproducing excessively and creating malignant tumors. We recognize such growths as "malign" by the single fact that they are no longer genetically identical to us. They have mutated into something else, something alien and not subject any more to the usual house rules. Something that needs to be seen as a threat and attacked, rejected instead of protected. They are "bad," at least in part, because they are in the "wrong" place.

This is one of the reasons for morning sickness and high blood pressure during early human pregnancy. These are signs of an early disagreement between mother and fetus, a fight in the womb between the parent and an outsider who shares only half of her genes. It is a dilemma faced and solved by all placental mammals, whose embryos "disguise" themselves by pretending to be maternal tissue; but avoided by others such as marsupials who lack the necessary chemical control. Opossums, for instance, expel their embryos when these are still too young to disagree, taking care of the rest of the infant growth outside, sometimes in a pouch.

Such problems of recognition are basic to all life and owe their existence to the fact that we carry in every body cell a reference library, of forty-six chromosomal volumes which hold the recipe for our kind of life. Instructions for making more of the same. Not the same organism, as we once believed, but the same genes because, as Oxford biologist Richard Dawkins has pointed out, bodies don't replicate themselves. Bodies grow and must, at a certain level, be seen as "survival machines," as vehicles for the true replicators, the genes themselves. Those "past masters of the survival arts," whose preservation and continuance is "the ultimate rationale for our existence."[101]

It goes against the human grain to think of ourselves, the paragon of animals, as rented and occupied in this way, as little more than taxis for bits of organic chemistry, but the evidence for such a view is strong and direct. We die, but genes don't. They never even grow old, but leap from body to body down through all the generations, manipulating each vehicle for their own ends, abandoning a long succession of used vehicles along the evolutionary highway, leaving each to rot there until another genetic passenger comes along and scavenges the spare parts to build itself a better gene machine.[406]

Our function in all this is not to reproduce ourselves. Until cloning becomes a realistic alternative, the best we can hope for is a part share in the future. My children are only half me, and my grandchildren no more than twenty-five percent. The mathematics is clear, individuality is fleeting and quickly diluted by sexual reproduction. But even if all this is in the best interests of diversity and natural selection, which keeps the genes going in a prudent variety of baskets, it would be wrong to underestimate the power of "blood." Dilution may dampen nepotism, but kinship still plays an extraordinarily large role in our lives and in our minds.

J.B.S. Haldane, the great British biometrician who saw everything from a delightfully different vantage point, was once accosted—allegedly in a bar by someone who objected to his Marxist leanings—and asked if he would give up his life for a brother. Haldane thought about this briefly and allowed that he would probably not, but he added: "I would sacrifice myself for *three* brothers or, failing that, *nine* cousins."

The reasoning behind this calculated response is simple. We share half our genes with brothers, one quarter with half-brothers and one eighth with cousins. The genetic imperative therefore is to save oneself until more than two brothers or eight cousins are at risk. Then it becomes mathematically worth the gamble of losing one's own life. It must be added however, in Haldane's defense, that on the two occasions when he was indeed called to save people from drowning, he did so without any actuarial thoughts because, "I had no time to make such calculations."[165]

It may not be necessary anyway for us to do these sums consciously in our heads. It seems that we are already programmed to behave as if we have and can, just as someone at bat in cricket or baseball manages, in no more than a fraction of a second, to complete all the differential equations necessary to compute a ball's flight characteristics and predict its trajectory. Mathematical ability and a knowledge of algebra have nothing to do with skill in such games. Good players do it all without thinking. They act instinctively, just as we all do when faced with a complex social situation that requires choices to be made. Some of these we make on the basis of direct experience or on the expectations of those around us, but a lot of the time we listen to a sort of whisper from within. There is something basic in us, and in most other species, which does take genetic factors into account in planning behavior.

As a child in Africa during the early 1940s, when adults seemed to be away a lot or otherwise preoccupied with world events, I spent a great deal of time amusing myself, much of it wandering on my own in the wild near our home. There was a wetland not far away, with the sort of seasonal pond that in South Africa is called a *vlei,* and along its margin every spring a pair of stilts nested. These are elegant black and white wading birds with long red legs that take readily to flight, trailing their conspicuous stilts behind them and giving a penetrating call. But when a few blotchy eggs have been laid in a shallow grassy depression, the behavior of the parents changes completely.

One of the pair I knew, usually the female, would remain on the ground, watching my approach and, when I got within about thirty feet of the nest, she would put on a performance worthy of an Oscar. The obviously healthy bird would suddenly become stricken by some sort of avian palsy. Her pointed wings began to jerk, her head droop, her legs turn to elastic. Stumbling and halt, clearly lame now in every limb, faint with fatigue, she staggered away along the shoreline, careful, I noticed, to avoid patches of thorn while picking out a suitable spot for her finale. The dying stilt puts swans to shame, agonizing and swooning, dragging out the poignant moment, hamming it up shamelessly, watching all the time to see if she has an appreciative audience.

I always was appreciative, coming back every day for more, and she never failed me. Carrying on with the bravura display almost at my feet, as I checked the eggs and later the chicks, following them right through to the moment of fledging. And this went on year after year, with the stilt risking her life, putting herself in danger, compelled, even though she must have known I meant her no harm, to go on doing so by a program not of her own making. One written instead by a calculating gene that sets the survival of four or five young stilts above her own individual safety.

There is no quarreling with the effectiveness of the display. Bird-watchers around the world have been charmed and distracted for ages by plovers, ducks, divers, owls, nightjars and other groundnesters that feign injury or illness to attract attention away from vulnerable eggs and chicks. It is hard not to admire such selflessness, which must, when faced with agile cats and foxes, quite often prove fatal. But it is even harder to understand how such behavior can have evolved within an evolutionary system that requires first of all that an individual act in ways that best enhance its own survival. Lying down in front of a predator is not likely to do that. It is suicidal, an altruistic act that seems only to benefit others. And altruism apparently has no place in Darwinian theory, or at least it did not until 1963 when William Hamilton, a young British biology student, rushed a paper into print.[167]

Hamilton was working towards his doctorate, as I was, at the University of London and we both submitted our theses at the same time that year. Mine was accepted and vanished forever into that archival maw which holds thousands of such studies designed only to meet academic requirements—foremost among which is that it should not upset one's elders. Hamilton's did and was rejected, but he had the sense and the courage to submit a brief outline of it to a more sympathetic American journal, and the following year published a fuller statement under the title of "The evolution of social behavior."[168] This, arguably, represents the most important single advance in evolutionary theory since Darwin's ideas were given genetic weight by the Austrian monk Gregor Mendel.

What Hamilton did was to argue that selection will naturally favor an individual's ability to discriminate between those around it on the basis of their kinship. He took the gene's eye view of association and produced some simple equations to show what conditions were necessary to ensure net genetic benefit from any interaction between individuals.

For instance, a foraging chimpanzee finds an abandoned bee hive with six perfectly good honeycombs. She knows from past experience that she can eat three of these herself, any more will make her ill. There are two other chimps working through the forest nearby and if she announces her find to them, there will only be enough honey for two combs apiece. What does she do? Give the chimpanzee "food call" and get to eat two, or keep quiet and enjoy the anticipated three delicious combs?

The selfish Darwinian chimp keeps quiet. The altruistic Hamiltonian chimp speaks up and looks good, apparently putting other interests before her own, affirming a growing feeling amongst ape-watchers that chimpanzees—especially the females—are actually very nice people at heart. Philopithecists. But the hard truth is that the altruistic act is probably nothing more than the result of a little self-centered genetic arithmetic.

It works like this. Honey is highly desirable. In the chimpanzee diet, out of a possible five points for nutrition, it rates a good four.

The net benefit to the selfish chimp who gets three combs is therefore 3 x 4 = 12. The altruistic chimp, after sharing her find, gets only two combs, a payoff of just 2 x 4 = 8. An apparent loss.

But one of the other two chimps is her brother, who shares half of her genes, so the net genetic profit to the family includes a further 2 x 4 x ½ = 4, bringing the total score once again to 12. Representing no apparent advantage or disadvantage to either pattern of behavior.

The balance in this transaction, however, lies with the third chimp, who happens to be a cousin or one-eighth relative, and his two-comb lunch adds 2 x 4 x ⅛ = 1 to the genetic scorecard, bringing the tally up to 13. This is an improvement on 12, and though the advantage is small, the conclusion becomes clear. Animals are systems which struggle to find a balance between energy output and input, and the chimp must choose the course of action which produces the largest net benefit, if not to herself, then to the genes she carries.

So she makes the call and shares the honey.

The net benefit is in the genes' favor. Nothing much happens to the genetic material in the individual who was faced with the choice. She was going to satisfy her sweet tooth either way and did not need to bother with any cost-benefit analysis. But what happens as a result of the choice she made is that the gene pool of which she is a part becomes just a little more likely to be filled with genes which influence bodies in such a way that these behave as if they *had made* such calculations. Genes work through us to help each other, and since Hamilton's insight, it has become necessary to look at everything we do with this in mind.

Haldane was right. The only time it makes biological sense to risk losing your life for another is when the person drowning is your identical twin. And even then it is a split decision. The arithmetic in this may be new, but the wisdom in it is universal. So much so that those who make awards for bravery, such as the Carnegie Medal or the George Cross, seldom if ever recognize courage involved in saving the life of a close relative. That much is pretty well expected. What is more unusual and therefore praiseworthy is putting yourself at risk to help a stranger. There is no genetic dividend in that. Genes seem to draw the line of their concern somewhere around third cousins, with whom we share only 1/128th of our heredity. Anyone beyond that is beyond the genetic pale, and not available for what has come to be called *kin selection.*

This exclusion is real and vital to the discussion of evil. Genes are essentially selfish except as far as close relatives are concerned. Under most circumstances, these alone are selected for special treatment. Everyone else is an outsider and fair game; and it is worth bearing this in mind as a basic rule in any analysis of behavior, even our own. There are several genetic instructions which seem to be common to, and appropriate to, all life. And Rule Number One among these is: BE NASTY TO OUTSIDERS.

Genes are simple-minded and mean-spirited. They have no vision and cannot be expected to have the welfare of the whole species at heart. Which is why universal love does not exist or make evolutionary sense. Generosity and unselfishness are not part of biological nature. Where such things exist, they have had to be learned or cultured by working against the trend. The sad fact is that we are born selfish.

So, of course, are all other living things—though it still comes as a surprise, the cause sometimes of rueful amusement, when we recognize our least attractive qualities in another species.

Penguins are perfect objects for such transference. Tuxedo-clad, vertical and clownish, gathered together in crowds at breeding colonies, they have a very human aspect. I never tire of watching them in action, enjoying their Chaplinesque gait and their straight-faced response to the sort of comic pratfalls that snow and ice so frequently occasion. My favorites are the Adelie penguins, who at a distance on an icefield, forming and reforming in social groups, are so easily mistaken for others of our own kind, promenading.

On one visit, I arrived at Hope Bay in the Antarctic in late summer, just as the crèches of young penguins there were becoming fully fledged and gathering in large numbers on the edge of the ice, waiting for their turn to go to sea. Adelies lead generally charmed lives, threatened by only one major predator—leopard seals, who make an easy living hanging around the colony picking off strays and inexperienced swimmers. A favorite tactic is to lurk beneath an ice floe or a rock shelf, lurching up and grabbing penguins just as they hit the water, before they can gather themselves together and shoot off beneath the surface at astonishing speed.

Learning to swim is a particularly dangerous time and I watched literally thousands of young birds crowding one section of coast, waiting to take their first plunge.

No one seemed ready to be a trendsetter and there was a great deal of anxious jostling on the edge, the sort of fake politeness that goes with the "No, please, after you" rituals in a dentist's waiting room. Those in the front ranks were being constantly replaced as penguins came forward, found themselves staring down into dark, unfriendly water and thought better of it, backpedaling and getting out of harm's way. Everyone waiting for someone else to find out whether there was a seal there or not.

The young birds were restless and hungry, anxious to get going and eventually the pressure became irresistible. Those in the back rows jostled and pushed until, inevitably, someone in the front slipped and fell into the water—and was immediately eaten. There was a swirl, a flash of sleek, spotted fur and gaping jaws, and the hapless penguin was gone. There *was* a leopard seal on the scene and he was hungry too.

Thousands of bright eyes saw the whole thing happen and there was an instantaneous change in crowd behavior. Where a moment ago there had been impatient scrimmaging and shoving, now everything was quiet. Every penguin there put its beak in the air, shuffled its feet and feigned elaborate disinterest in the whole endeavor, looking as though swimming were the last thing on its mind. Peace reigned, but the pantomime didn't last long. Within minutes, the mood changed again, impatience reappeared and the whole cycle started once more, with exactly the same result—another meal for the leopard seal.

I watched for almost an hour and was glad I had, because the finale was well worth waiting for. It came after the fourth or fifth sacrifice, with yet another penguin coerced into being "it," the guinea pig. Once again all eyes focused on the splashdown, but this time there was a difference. The young "volunteer" submerged briefly and came back to the surface intact, swimming round in circles, testing the paddle-power in his wings, tasting the cold salty water. He kept on doing this for perhaps thirty seconds, oblivious to the largest audience in the Antarctic. They peered at him, watching intently, looking left and right to make certain that the delay was real and not just some cynical new seal maneuver, and then, as though the all-clear had been sounded, plunged as one into the water, filling Hope Bay with the sort of splash and clamor I hadn't seen since my last school swimming gala.

I make only the most perfunctory apology for this diversion. I enjoy penguins and they provide an excellent example of selfishness, perhaps even cowardice, in action. It is easy to sympathize with their situation and to identify with their response. The dilemma is real and not uncommon. We frequently find ourselves in situations where it seems most prudent to let someone else take the lead, admiring their courage in doing so, admonishing such heroics only when they involve close kin, someone with a significant fraction of our genes in their brave bodies. Then the behavior seems suddenly reckless and foolhardy, enough so to make us angry, and then to wonder why anger should be an appropriate response.

For the genes it is. Kindness to strangers is not part of their plan. Such behavior, on their balance sheets, means wasted energy and unnecessary risk. So we find ourselves torn between courage and fear, hate and love, and end up feeling guilty. And that is the next stage in behavioral analysis—a recognition that genes are not impartial. In addition to being generally nasty to outsiders, genes have a supplementary tendency which could be described as Rule Number Two: BE NICE TO INSIDERS.

Hamilton's genetic theory has made the subject of "niceness," of altruism, central to evolutionary studies since 1964 and spawned a whole discipline called sociobiology. The most significant contribution of this still controversial area of study has been to confirm that no organism lives for itself. It is, in Edward Wilson's words: "A unique, accidental subset of all the genes constituting the species."[425] A vehicle giving scientific substance to Samuel Butler's famous aphorism, that the chicken is just an egg's way of making another egg. An animal or plant is just the gene's way of finding subsidized housing.

Dawkins and Hamilton and Wilson all recognize "the gene" as a biochemical entity, a collection of DNA molecules arranged in a particular fashion and housed along ribbons inside the cell nucleus of most higher organisms. This is fine, as far as it goes, but it is worth bearing in mind that some influences which direct our evolution and development might well lie outside the cell nucleus, or even outside the cells themselves. And that, while I join with the three wise men of sociobiology in a sort of general genetic determinism, I am more often using "the gene" simply as a convenient metaphor for all the forces which govern instinct and the inheritance of both form and function in any individual.

In evolutionary time, the individual organism counts for almost nothing. Yet it exists and behaves in ways we can watch and measure. Ways which often suggest that it knows, or has been programmed to make it look as if it knows, which patterns of behavior will most benefit its own genes. Very often these are against the best interests of the individual, who may even be called upon to give up her life for a large enough number of relatives to satisfy the gene's arithmetic. And surprisingly often, the decision goes against the interests of the species, simply because selection does not take place at species level. Genes care nothing for "the good of the species." All they are concerned with is becoming better represented in the gene pool, and in this endeavor they are ruthless.

As a biologist, and even before we have a full understanding of how genes work, I find it useful to think of "the gene" as an entity. One without direction or intelligence, but one otherwise equipped with everything necessary to provide a framework for the origin and evolution of evil.

There seem to me to be two necessary preconditions for becoming wicked—and these are the ones still recognized in every court of law. There must be both *means* and *motive.*

The Principles of Pathics provide the means. Good things, as we have seen, become bad as a result of loss of place, loss of balance and loss of diversity. But for such disarray to persist in any meaningful way, there needs also to be an influence which gives the process momentum. There needs to be a motive, which is not necessarily conscious or creative, but has to have survival value. And this is provided in full by genetic "selfishness," by the effect the first two Genetic Rules have of creating an essential divide between Us and Them. A schism which has nothing to do with "the good of the species." All genes are concerned with is becoming better represented in the gene pool, and in this endeavor they can be ruthless.

As a student at London Zoo and later as director of a large zoo in Africa, I had trouble with hyenas. The sexes in spotted hyenas are so much alike that the only way you can be certain that you do in fact have a pair is if they breed. Female hyenas have a clitoris the same size as the male penis and equally erectile. Her vaginal labia are fused to form a pseudoscrotum which is filled with fatty bodies as large as male testicles. In every external respect the sexes are identical and when two hyenas meet and go through a greeting display that involves mutual sniffing and licking, both male and female genitals become erect and the animals stand head to tail with one hind leg lifted in precisely the same way. To a human observer there is no apparent difference between them, even when one has a hyena close enough and compliant enough to handle. Which leads to real confusion for zoo people, who begin to suspect that this may be what gives "laughing" hyenas cause for such demented nocturnal giggling.

Hyenas do of course mate and bear young, so they must be able to discriminate between their sexes—probably by smell and at least during the time a female is in estrus. Female hyenas are in fact excellent mothers, doing what has to be done for successful reproduction, but during the remainder of their lives they seem to operate in thrall to a program which has little to do with hyenas and everything to do with their selfish genes.

Spotted hyenas lead a complex social life in which females play the dominant role. Clans of as many as seventy or eighty individuals are always led by females, and all adult females are dominant to all adult males, who leave the group when they reach puberty. Females assert their dominance by being far more aggressive than the males at every stage in their lives, producing one of the most chilling phenomena ever seen in any mammal.

Hyenas usually have twins, which are well developed at birth, completely furred, eyes open, canine and incisor teeth fully erupted. This is true of some other carnivores, but what makes the spotted hyena exceptional is that within minutes of birth one of the cubs attacks its twin, sometimes savaging a brother or sister that has not yet even emerged from its amniotic sac. If both offspring survive this preemptive strike, the battle begins in earnest as "the pair roll in a bitter embrace, each with the skin of the other locked in its jaws."[131]

This is not the normal rough and tumble of siblings at play. This is a fight to the death between two baby animals scarcely an hour old. There is nothing unusual about sibling rivalry. Offspring routinely compete with each other for food and attention. I have seen heron fledglings struggle fiercely with each other for every handout from a parent returning to the nest; and once, over an agonizing week, watched one of a pair of black eagle chicks die as the other progressively pecked it to death. But these examples were the result of food shortages and demonstrate a healthy and natural tendency for the fittest to survive. What is disturbing about the hyena neonatal conflict is that it seems to be the norm. Spotted hyenas nearly always have twins and seldom get to raise more than a single cub, because the cubs themselves seem to be genetically programmed to attack and kill their siblings on sight.

This is a rare example of pure aggression, taking place without any preliminaries, independent of the sort of social history that usually leads to conflict. Even by the gene's pitiless standards, this is strong stuff. Fratricide or siblicide has a long human tradition, playing a prominent role in family tragedies from Cain and Abel to *East of Eden.* But it is not common in nature because it carries a heavy penalty, the certain loss of at least half your own genes. The fact that spotted hyenas do so routinely suggests that this loss is one worth taking. It must be offset by an even larger gain elsewhere. But where, exactly?

The answer to this mystery may lie in discoveries made by Laurence Frank working on wild hyenas in Kenya and on a group of captive animals in California. Over a period of fifteen years, Frank has found that the cub which usually wins the murderous battle in the den is female, and that if she is the daughter of a top-ranked female, she will invariably replace her mother as head of the clan. This is clearly something the genes would like, the kind of continuity they thrive on and seem to have contrived by a very simple chemical device. With a series of blood tests, Frank has also been able to show that androgens run riot in the whole population. Spotted hyena fetuses, no matter what their sex, have levels of testosterone higher even than adult males. Something has switched on the process which produces hormones that stimulate the development of masculine characteristics, and done so early and indiscriminately, right there in the womb, leading to conspicuous later changes in body, brain and behavior. And the result is bizarre genitalia, sexual confusion, and adult females so large, competent and aggressive that they run a tight and genetically stable matriarchy.[131]

The only saving grace in this frightening story of genetic imperialism lies in Laurence Frank's additional suggestion that hyena mothers might sometimes intervene in fights between their offspring to tailor the sex ratio of the community. He reports on a depleted clan which badly needed replenishment suddenly allowing an unprecedented number of pairs of female twins to survive. A subtle and heartening reassertion of individual independence over genetic hegemony. Perhaps. It is not always easy to tell where the orders are coming from. Unscrupulous genes are also inclined to cheat.

It takes an enormous amount of energy and effort to rear young. The advantage in evolution often goes to those species which take such responsibility most seriously, protecting and caring for their offspring until these are able to take care of themselves. But there are obvious benefits to be derived from any ploy which can persuade someone else, a childminder, to take over these chores.[119] Some social species—such as white-fronted bee-eaters in Africa, almost half of whom forgo breeding in any year to assist their fellows to feed and rear fledglings—share the responsibility, rearing their young cooperatively. But a few have passed the whole affair over to another species altogether, and to do so, they cheat.

Most renowned among these dissemblers are the cuckoos, more than fifty species of which lay their eggs in other birds' nests. They intimidate their unwilling hosts into leaving the nest unguarded by imitating a local predator, such as the sparrowhawk. Or the males may distract a host's attention while the female sneaks in and lays eggs which mimic those of their habitual host in size and color. Or they simply destroy the host's clutch of eggs altogether. Cuckoo chicks also bear a close resemblance to those of the host species, even copying the distinctive color patterns which act as food guides in a gaping chick's mouth. Everything possible is contrived to persuade the hosts that they are hatching and rearing their own offspring, but if all else fails, the cuckoo chick takes matters into its own wings and simply eliminates rival eggs or nestlings by throwing them out of the nest.

Some baby cuckoos are programmed to do this. If anything presses on their backs during the first day or two of their lives outside the egg, they climb backward up the side of the nest, supporting this burden between their wing stubs, and heave it overboard. This pattern of behavior is instinctive, gene-controlled, part of the normal repertoire of cuckoo behavior, fixed there by millions of years of freeloading. It is what we expect of cuckoos, and only cuckoos. But in 1973 three young Spanish biologists made a discovery that has forced us all to think about such assumptions again.

They were working in the Donana Reserve, trying to find out how magpies deal with brood parasitism by the local cuckoos. They introduced a variety of egg models and eggs from other species into magpie nests and found, as expected, that magpies most strongly reject eggs which look least like their own. They also tried the featherless chicks of other birds, and found that few of these were rejected. Again not surprising, since it is eggs and not chicks that parasitic cuckoos deposit, so there has never been any need for magpies to develop an anti-chick device. In any case, cuckoos have anticipated them in this too by producing chicks that mimic the appearance and behavior of the magpie's own brood. However, one of the chicks forced on the magpies by the researchers did something quite extraordinary.

It was a European swallow chick, a species that is not parasitised by cuckoos. It was put into a magpie nest already occupied by three magpie eggs. The next day the researchers noticed that one of the eggs had fallen to the ground below the nest. It was not damaged, so they picked it up, replaced it, and watched. "We were able to observe that the swallow chick repeatedly loaded an egg on its back whilst climbing on to the edge of the nest, dropping the egg on the ground below the nest twice in front of our eyes."[5]

There is revealed, even in this bald statement, a little of the incredulity which all biologists must feel about this kind of event. It goes against everything we think we know about how evolution works. The swallow chick, normally weak and helpless, was behaving just like a cuckoo. And the problem is that it is nothing like, nor even distantly related to, cuckoos. Nor does its species have any direct experience of cuckoos. Baby swallows are not accustomed to finding themselves in magpie nests, or any nest other than their own. And yet, when a swallow finds itself in this predicament, faced—as far as we know, for the first time in history—with the problem of having to compete with eggs and nestlings larger than its own, it performs a difficult and complex task to perfection.

Only the genes could orchestrate such a response. There is nothing in the normal life of this species of swallow that could justify inclusion in its own gene pool of a dormant anticuckoo adaptation, something preemptive, just in case. Evolution is usually very frugal about such things. The only explanation that makes sense is one proffered by that champion of the selfish gene, by Richard Dawkins himself, and has nothing to do with cuckoos at all. It has to do with swallows and with the possibility that this may be what baby swallows sometimes do to each other. After all, he says: "the ruthless behavior of a baby cuckoo is only an extreme case of what must go on in any family. Full brothers are more closely related to each other than a baby cuckoo is to its foster brothers, but the difference is only a matter of degree."[101]

Perhaps we are all capable of the sort of simple arithmetic that leads to fratricide. Sometimes it takes a silly question, the posing of an unlikely problem, to elicit the most unexpected response from an animal.

European swallows normally lay four eggs in a clutch. This is the optimal number as far as the species is concerned, but a baby swallow might see things differently. As far as it is concerned, a smaller number would be better, provided that he is one of them. Unlike a cuckoo, who has no genetic interest in the rest of the nest, a swallow might not want to destroy a whole brood of his brothers and sisters. But he could benefit directly, as an individual, by cutting down the size of the clutch. Tipping out one egg would give him an extra quarter share of the parental care, and that sort of advantage is not something individuals are slow to exploit. When it comes right down to it, there is a little of the hyena in each of us.

Many species make allowance for risks to their offspring by producing more than they need to do for simple population replacement. Birds whose livelihood depends on irregular or unpredictable food supplies are particularly vulnerable at nesting time and their genes seem to have dealt with this dilemma in a distinctly draconian fashion, encouraging not only fratricide, but also infanticide and suicide.[291]

The mathematics involved is equally unforgiving. When food is short, no nestling in a brood gets enough to eat and all die. But if one of the chicks is removed in some way, all the others stand a better chance of surviving. Their individual fitness is improved. But because Hamilton's equations are concerned with "inclusive fitness"—the total chance of reproductive success—they come to the hard, but biologically sound conclusion that each nestling can attain greater fitness by dying than it can through continuing to live. The chips are stacked against individuals right from the start. All that remains to be decided is which chick should be sacrificed and when.

Ornithologists have long been familiar with brood reduction in a variety of species and have generally assumed that this was under simple parental control. That when food was short, they would feed the nestlings selectively and allow one to starve—usually the last one to hatch. Sometimes this is exactly what happens. But it is important to appreciate that the surviving chicks benefit from brood reduction far sooner than adults. Recognition that siblings and parents may have different interests has led to a search for other strategies and found that siblings sometimes preempt parental choice by turning on one of their own number, pushing it out of the nest, pecking it to death, eating it or simply preventing it from being fed. Fratricide seems to precede infanticide, particularly in species with small broods, where the selective advantage to the surviving chicks is greatest.[26]

The most surprising discovery, however, has been that a third strategy also comes into play. In addition to fratricide and infanticide, one of the chicks—usually the nestling with the shortest life expectancy—may choose to commit suicide. This most often involves leaving the nest before it is capable of surviving on its own, though it is difficult to separate true Oatesian self-sacrifice of the "I am just going outside, and may be some time" sort, from the simple avoidance of sibling and parent pressure in the nest.

As far as the genes are concerned, suicide can make sense. The runt of the litter, the backward chick in a nest, is often just the parents' way of hedging their bets. If food is abundant, the extra mouth can be fed. If food is short, losses can be cut before too much time or energy is invested. Darwinian struggle for survival requires that the runt does everything possible to live, but genetic theory and actual practice suggest that, more often than not, it goes gracefully, perhaps even willingly, contributing directly to inclusive fitness by being eaten, or indirectly by enhancing a relative's chances of survival.

There is even a case to be made for such selfishness in the womb. Natural miscarriages are common in our species, perhaps ending as many as one in every five pregnancies, and are usually described as "spontaneous." That comforting word means, more often than not, that we just don't know what happened. But such loss could be an adaptive strategy. If the embryo involved is damaged or untimely, representing a bad investment on the part of the mother and her genes, it would be best for all concerned if she aborted. Quite often she does. A study of female prison inmates in the United States produced the surprising finding that, despite their history of prostitution, drug addiction and other life-threatening lifestyles, they were seven times less likely to give birth to children with defects. This counterintuitive statistic is however balanced and possibly explained by the fact that they were also three times as likely as those mothers outside prison to have had at least one stillbirth or natural abortion.[35]

It seems that "spontaneous" abortion is more likely if the mother is leading a stressful life, and that when such miscarriages occur they tend to discriminate against less competent fetuses. Against the "runts" and those carrying birth defects. It is possible that the mother may be involved in unconsciously denying such an unwelcome embryo the necessary nutrition, just as a parent bird discriminates against a surplus nestling; or that she may expel it prematurely, just as parents and siblings ease a backward chick out of the nest. But it is also possible that a fetus may cooperate in its own abortion, committing early suicide if you like, because such action is likely to contribute to the fitness of the mother. Embryonic altruism of this kind could be widespread, but it still needs to be recognized for what it is. Another case of cold-hearted calculation by genes, which reflects the fact that a fetus shares half of its genes with its mother and can expect to share half with the mother's next child. By giving up its own life, the altruistic embryo confers a two-fold benefit on the next reproduction—and increases its own inclusive fitness. The genes win again, this time fairly, but that doesn't prevent them from taking unfair advantage whenever possible. In fact, you can count on it.

In addition to being generally nasty to strangers, and nice to relatives and friends, genes are congenital crooks. They hedge their bets by a loophole law which may be described as Rule Number Three: CHEAT WHENEVER POSSIBLE.

Cheating comes naturally. The genes demand it and children soon learn that it pays. Hence the battle of the generations. No chick, no pup, no baby ever passes up an opportunity to deceive. They pretend to be younger, hungrier and more at risk than they really are, in the hope of getting more time, more food or more attention. I have seen pelican chicks in Baja California produce what looks like convulsions, flinging themselves about the nest, biting their own wings, literally "throwing a fit" in frenzied attempts to gain advantage at feeding time. It is always a game of nerves, a psychological conflict in which lying, cheating and blackmail are the major weapons, used unscrupulously to exploit parental concern and guilt.

Some cuckoo chicks, for instance, have the very nasty habit of screaming so loudly for food that foster parents feed them first lest their din attract a predator. And if you think that comparisons of this kind of habit with human behavior are far-fetched, I offer the example of Lynn Kivi, a young mother in Woodstock, Georgia, whose patience snapped in 1994 when her fractious son wouldn't stop whining for something he wanted in a supermarket. She tried to keep him quiet, but he exploited the advantage of being in a public place to such an extent that she finally slapped him. Unfortunately for her, there was a "predator" on hand and Mrs. Kivi ended up in handcuffs, arrested on charges of cruelty to children, facing a possible twenty-year prison sentence.[16]

Such official retribution in human society is rare, partly because we condone a certain amount of discipline, even physical punishment, as a parental right. But we are becoming more sensitive to the possibility of traditional discipline escalating into genuine child abuse and may now, as a result, overcompensate. It has never been easy anyway to assess such social situations, because parents and their offspring have imperfect knowledge of each other. A parent never knows exactly how much care an infant needs, and the infant never knows how much attention a parent can afford to give. The odds, however, are usually stacked in the child's favor, because no parent dare take the risk of assuming that the child is not genuinely hungry, sick or frightened. And no offspring anywhere miss the opportunity to take advantage of indecision, exploiting such situations until they "cry wolf" once too often and parents get wise. Then the balance shifts, generosity becomes leavened with tough-mindedness and the response is more likely to be disengagement or disinterest. "Not now, dear, I'm busy."[28]

Domestic and social behaviors are all highly political, filled with complex strategies and counterstratagems, but we are beginning to understand how these situations arise and where the initiative really lies. Hamilton has helped enormously with his concept of inclusive fitness—the sum of an individual's own fitness, plus the sum of all the effects such fitness has on his relatives. With such arithmetic, we can begin to see how simple acts can have very different results, depending upon the relationship of those involved.

For instance, if I give up food or shelter or even a mate for my brother's sake, I diminish my own fitness. This is *altruism,* a sacrifice that seems to lead only to genetic loss. But it can also put him into a position where he is able to reproduce successfully and pass more of the genes we both share, including perhaps a tendency to be altruistic, on to the next generation. On the other hand, if I refuse to make sacrifices for the sake of the family and turn down my brother when he comes to me for a loan, I diminish his fitness. This is *selfishness,* and does him no good, but it may however pay off in the long run if it increases my fitness to a point where I can have more, possibly equally selfish, offspring. Genes find both strategies effective and make no moral judgments between them, using either to their advantage, depending on the situation. We must expect to find unthinking generosity and greed, altruism and selfishness, in the behavior of every species.

As a child in Africa, I remember being taken to watch a springbuck migration in action. At that time, tens of thousands of these small antelope with short curved horns still moved across the edge of the Kalahari Desert, following the seasonal rains. These represented only a fraction of the great herds which once traveled that way, but it was nevertheless an astonishing sight. Their warm brown bodies flowed through the arid landscape like a river of sand, dividing to pass around a clump of thorn on a rocky outcrop, rejoining on the other side. It was like watching the whole desert move, but every now and then the giant organism would splinter as something startled one of the buck and it would leap six or eight feet vertically into the air, flashing a dramatic black and white flag on its side, paddling with its hind feet, raising a crest of snow-white hair on its rump until this shone like a light.

It didn't occur to me at the time, but this *pronking* (from the Afrikaans for "showing off") is not just a dramatic display of high spirits and tight sinews. It is an altruistic act. The law of successful preydom is "Don't make yourself conspicuous," and yet here was an animal in territory frequented by lion, leopard, cheetah, hyena and hunting dog making an exhibition of itself, inviting attack. There is some safety in numbers, certainly, but equally compelling evidence to suggest that many predators single out an individual that catches their eye and hunt it down deliberately, to the exclusion of others in the group. Pronking attracts attention, it held me in thrall even as it put the leaper at individual risk, but it is also an alarm call, alerting other springbuck to the presence of danger, preparing the whole group for evasive action. It hasn't, unfortunately, helped them against human predators. Springbuck have been hunted now almost to vanishing point, but it is possible in some protected areas to see small groups of these elegant antelope, and they still pronk. The ostentatious habit remains because, on balance, the genes are very canny accountants and they find it profitable.[403]

In sardines, anchovies and other "bait fish," the opposite situation appears to prevail. These species gather in groups, they school, for completely selfish reasons. Small fish are governed by their instincts to seek cover and, in absence of convenient rocks and reefs, they find shelter in a crowd. Each fish behaves in such a gathering as if it were trying to put other fish between itself and a potential predator. And the net result of such outright selfishness is the familiar centripetal school, which swirls and reshapes itself, over and over again, as each individual tries to move from the perilous edge to the more secure center of this association.[422] The selfish school is more conspicuous, less maneuverable than a lone fish, and predators get their meals more easily. The average fitness in the school is lower than it would be if the fish did not school. But they have no choice. In this case, the gene for selfishness prevails, at least for now, and the inevitable consequence of everyone running for cover behind everyone else, is that they soon become bunched together in a school. This association at least may not be the product of finely tuned ecological checks and balances, but the unavoidable outcome of purely mechanical forces that sometimes may turn out to be bad for the species.[170]

The genes, however, don't care. They never have, concentrating single-mindedly on their own agendas, lying, cheating and even killing where it serves their purpose.

In addition to altruism and selfishness, there is another permutation possible for genes interested in improving their profit-loss ratio. Going back to the simple relationship between a pair of brothers, it is possible for me to help my brother indirectly by harming an unrelated competitor. I gain nothing myself. In fact, I might even lose, but go ahead anyway, perhaps in the hope of increasing the fitness of my brother. This strange pattern of behavior is called *spite.* It is all too human, largely because it depends on complex awareness of bloodlines and a capacity for intrigue, but it would be wrong to assume that only we can be spiteful.

The question is, could we expect an animal to harm itself merely in order to harm another more? And the answer seems to be, certainly. When choosing a behavioral strategy, an animal can increase its inclusive fitness by being nice to close relatives, indifferent to distant relatives, and antagonistic to strangers. If such antagonism takes little time or effort, if it costs the animal nothing, we can expect it to do a lot of it. But if the behavior is difficult or expensive, we can also expect him to be less quick to act, more deliberately and indiscriminately spiteful.[169]

American marsh wrens in an area occupied by a variety of other songbirds have been seen entering blackbird nests and pecking away at their eggs until these crack or puncture, killing the chick inside.[294] Australian bowerbirds often saunter over to the elaborate structures in which neighboring males display and mate, and comprehensively wreck these in the absence of their owners.[263] Elephant seals on their crowded breeding beaches frequently fall prey to apparently gratuitous aggression, with adult females even ganging up to murder someone else's pup.[239]

There is a lot of spite about, some of it perhaps just selfish in the sense that it does present some selective advantage to the actor, who succeeds as a result in gaining more food, more space, greater status or increased reproductive opportunity. Even the murderous elephant seals may profit from a reduction of competition in the lives of their own offspring, but pure spite still has a place in evolutionary practice. Anything I can do to help my brother is good, and anything I can do to hinder competing strangers could be beneficial to both of us. I suspect that the reason genes are so tolerant of spiteful behavior, even where it involves a considerable investment of time and energy, is that it suits their selfish purposes. If nothing else, it tends to reduce the possibility of random mating and genetic dilution. We prefer to stick to those most like us. Would you let your daughter marry a spiteful man? Perhaps. Because spitefulness is at least out in the open. It doesn't pretend to be anything else. More difficult to deal with or assess are those behaviors which are intended to deceive. And the best known of these are lies, which can usefully be defined as "the communication of misinformation to members of your own species." Rule Number Two requires us to be nice to insiders, so we frown on deceit where these are concerned. But Rule Number One demands that we be nasty to outsiders, so lying to them is perfectly acceptable. They are fair game. Everybody knows that.

Take mimicry for example. Functionally, it represents a sort of active camouflage, an attempt not just to fade into the background, but to look like something else altogether. Invertebrates are very good at such deception. In Madagascar, I have seen a group of brightly colored flatid bugs clinging together on a twig in perfect imitation of a pink spike of flowers; and green wasps clustered on their hanging nest so still, even at my approach, that I took them for a knobbly fruit common in the area and very nearly picked it. In the same dry forest there are predatory jumping spiders masquerading as wasps, and larger orb web spiders shaped like dead leaves hanging innocuously near the center of each gossamer ambush. There are longhorn beetles in this garden of deceit, artfully arranged to look exactly like the splat of a bird dropping; and stink bug nymphs that gather round their old egg cases, waving their legs in collective and very persuasive imitation of a poisonous caterpillar. All these statements—I am a flower; I am a fruit; I am not a spider; I am a leaf; I am dead; and I am poisonous—are untrue. They are calculated to mislead and misinform other species that prey on, or are preyed on by, the ones sending these creative messages.

In the Amazon such invention runs rife, piling deception upon deception. There is a planthopper, a bug related to the aphids, whose head has become enormously bulbous and elongated, drawn out into a fake snout with large eye bumps behind, each of which has a white mark in exactly the right place to simulate the light reflected from a vertebrate eye. Along the sides of this "snout" are grooves that make it look like a partly opened mouth, and staggered along these are a series of apparently bloodstained false teeth, rendered not only in color but in bas-relief. The total effect is an incredibly lifelike imitation of an alligator, small, but perfect in every detail. I have often come across these heavily disguised planthoppers dozing on the banks of muddy tributaries and, despite the fact that I know them for what they are, and have myself little to fear anyway from a four-inch alligator, I find it takes a conscious effort of will to grasp one by the head. And even when I do, I have to be prepared and steel myself for a further sneaky line of misinformation. When alarmed, this duplicitous insect opens its wings, revealing in a flash two large eye-spots that stare right back at you, exactly like the face of an angry horned owl.

I find it hard to believe that this astonishing array of duplicity can be the product of sheer chance. The effect is so realistic, so dramatic and so perfectly calculated to be so, that the mechanism involved cannot have been the planthopper's alone. It was certainly a coproduction involving predators, probably birds such as herons and flycatchers, who came along at various stages of the long evolutionary series, acting as objective art critics, selectively rejecting the imperfect works, removing these from the gene pool by eating them. But it is difficult to imagine any way in which such elaborate designs could have been perfected in the time available without some direct reference to the owl or the alligator. Some sort of elaborate flux of information between all the species in that part of the forest. An interchange of pattern and instruction which gives normal Darwinian natural selection something concrete to work on. I suggest that misinformation of this kind, which involves such meaningful, appropriate, inventive, evolutionary adaptation, requires the existence of an influence which goes well beyond the bounds or the abilities of the individual or the species.

And I can think of no system better designed to provide such information than the genes themselves, those canny strategists whose being can be as subtle as the code carried in a virus that flits from body to body spreading news like gossip at a cocktail party.

How else can one account for communication not just across species lines, but even between kingdoms? There is an orchid in Algeria whose small flowers shine like blue-steel looking-glass, edged with gold and set in velvet. These are unlikely patterns in a plant, but put together they bear an uncanny resemblance to the female of a local species of wasp, sitting with her half-crossed wings reflecting the blue North African sky. The thick set of red hairs on the insect's abdomen are imitated to perfection by similar vegetable bristles, and the antennae of the wasp are beautifully reproduced by the threadlike upper petals of the orchid. All in aid of attracting a male wasp who tries to copulate with the apparently receptive female and ends up with pollen on his face.[234]

While misinformation of this kind is common between species, one might expect members of the same species to be more honest with each other, revealing factual information about things such as sex, readiness, whereabouts and the limits of an established territory. Far from it. No system of communication in any species is designed to convey the truth, the whole truth and nothing but the truth. Signals may be selected to carry correct information, but they are equally likely to pass on misinformation or a delicately balanced mixture of the two. When a male scorpion-fly, equipped with the customary nuptial gift of an edible insect, is faced with what looks like a receptive female, he has to decide whether she is indeed a potential mate worth courting, or whether she is a rival male merely pretending to be a female and intent instead on stealing the wedding present. And a female, faced with what looks like a suitable suitor, has the equally difficult task of deciding whether his intentions are honorable, or whether he is one of those unscrupulous scorpion-fly males who has simply secreted an insect-shaped gob of saliva in the hope of deceiving her long enough to mate. It is a sad truth that, throughout the animal kingdom, any individual who can cheat in order to gain an unfair advantage will do so.[374]

It pays to lie, and if you are going to do so, nothing works quite so well as the "big lie."[402] Political propaganda is often well founded in evolutionary theory. The investment of time and energy in a variety of deceptions is likely to differ little and, if the chances of a small lie being believed are no more than 50:50, you might as well go for broke with gross misrepresentations of the truth. Making spots on your wings is hard work, but while you are at it there is more to be gained by making them look like an owl's eyes than simply pretending to be another kind of insect. The message: "I am not a crook," or "I am not the bug you think I am," is not nearly as effective as: "Watch out for card-carrying Communists," or, more convincingly, "Beware of the Owl!"

Dian Fossey once watched a gorilla group being led along a mountain trail by an adult female, who noticed a clump of flowering vine almost obscured in a tree. These are a special treat for gorillas, so, without looking at those behind her, she sat down beside the trail and became apparently absorbed in self-grooming. The group passed her by, but as soon as the last one had gone out of sight she climbed quickly into the tree, consumed the flowers and then rushed off to catch up with her party.[129] Such tactical deception is common amongst primates. This example was one that involved deliberate concealment, a sort of misdirection by default, which resulted in profit for the actor. But there are many more complex examples in which an actor actively misleads other individuals, manipulating both their attention and their sociality to its own advantage.

One of the most vivid comes from a field study of chacma baboons in Africa by Richard Byrne and Andrew Whiten.[58] Their troop was being watched at a critical time, when an older male was being replaced by a younger and more vigorous rival who monopolized the mating females. But baboons cover large territories whose extent and potential are best known and understood only by older members, who still tend to lead the way when the group moves. On this occasion, the deposed leader had found a new and rich feeding ground near the edge of their home range, but before they could exploit it, the brash young leader muscled his way into the center of things and reasserted his authority. Instead of disputing this move, the veteran mounted an elaborate deception. He rose suddenly, leaving the site and setting off briskly in an obvious direction. The others milled around uncertainly for a while, but within minutes all of them, including the young tyrant, followed his initiative—as they had become used to doing. Two minutes later, the wise old baboon reappeared on the site, having swung round in a circle and settled down to enjoy the rich food source without interruption or competition.

Lying is merely the most obvious way of cheating in a social situation, and so common that it must be seen as universal. A part of our environment as natural as wind and water. Yet it only exists because "honesty" already forms part of the same system. Deception cannot work unless it is a *rare* variant of an honest response, unless it mimics a preexisting pattern of appearance or behavior and exploits the tendency of most individuals to accept that at face value. In close-knit groups of animals, where individuals recognize each other and interact with each other often, deceit is harder to practice and easier to detect. So it has had to become even more subtle, and our sensitivity to it ever more acute. The three rules of genetic action are ground rules only, enough to get the game of association going, but success in it depends in the end on the sort of fine-tuning that makes a complex society possible.

Social intelligence is given far too little value in societies which measure worth by standards of technical prowess. Even our attempts to rate the relative intelligence of other species have been hampered by experiments which assess their ability to deal with physical gadgetry, rather than with each other. What makes the social environment so special and so challenging is that it is reactive. It talks back and requires a far more sophisticated approach to problem-solving. It needs, and probably has encouraged, the kind of social expertise which has come to be described as "Machiavellian," after the infamous Florentine who highlighted the value of deception and manipulation in social and political affairs. "For a prince," he said, "it is not necessary to have all the virtuous qualities, but it is very necessary to appear to have them . . . and yet also to be prepared in mind that if you need to be not so, you can and do change to the contrary."[255]

This kind of social intelligence preceded technical intelligence in evolution. It is even possible, as psychologist Nicholas Humphrey has suggested, that creative intellect arose as a direct result of having each other to manipulate and explore. Life for mountain gorillas in Rwanda or chimpanzees in the Congo forests is relatively easy. Food is abundant and easy to harvest, predators are few and easy to avoid. There is little to do but eat, sleep and play. But the calm is illusory, it looks easy because the members of such groups have discovered how to create and maintain stable societies. They may be physically lazy, but they are intellectually hyperactive. The forest may not present great problems to them, but other gorillas and chimps certainly do. Social survival requires the use of every skill they or we can muster, largely because such society is a paradoxical concept. On one hand, it provides benefits for those individuals who preserve its structure, who keep the group together. But on the other hand, it encourages and rewards those individuals who discover how to exploit and outmaneuver their fellows. By its very nature, a complex society creates calculating beings—ones who recognize the consequences of their own behavior, who predict the response of others, and who measure the net profit and loss in everything that happens.[202]

Add to this the genetic pressures created by conflicts of interest between the generations in any breeding community, as well as the possibility of competition with other breeding lines, and you have the makings of a system in which selfishness is guaranteed. Any heritable trait which increases the ability of an individual to outwit others will soon spread through the entire gene pool. It is clearly rampant in ours and has given rise to the playing of extraordinary Machiavellian games.

At the heart of all these lies one necessary assumption, a basic one that unites all three genetic rules. And the gist of this is that genes will do whatever it takes to protect themselves, even if it means sacrificing some of their own.

Langurs are elegant, long-tailed monkeys with eyebrows that are linked in permanent surprise. They range through India up to the foothills of the Himalayas in troops of as many as seventy or eighty animals, always including just one breeding male. Other males leave or are driven out of the troop and form bachelor bands which shadow such troops, monitoring their strength, looking for signs of weakness, waiting for an opportunity to usurp the position of the resident male.

Competition for his position is fierce and may involve the adult females, all of whom are closely related, joining forces to drive invading males away. But every two or three years the interlopers win and one of their number defeats the incumbent male. When such displacement is successful, it is the signal for a reign of terror in which the new leader kills dependent baby langurs in the troop. Females fight back against the murder of their young, but in the end their bodies betray them, coming back into estrus again within days of the loss of their infants. And the frenzy of infanticide ends only after seven months, when females who have become pregnant by the usurper begin to produce young that carry his own genes instead of those of his predecessor.[196]

The genetic imperative here is obvious and brutal, but it is not simple. The langur female's reproductive success is affected by the murders and *her* genes encourage countermeasures which include cooperative resistance to the killings; leaving the group to follow the old male who fathered her young; or even living dangerously on her own. But most impressive of these strategies is one which involves a pregnant female "pretending" to be in estrus, even to the extent of appropriate hormonal changes likely to convince the new male of her readiness, and copulating with him often enough to allay his suspicions when her infant is born long before the normal period of pregnancy has elapsed. It usually works, persuading the despot to accept the defeated male's infant as his own.[197]

Tactics of this kind may not be conscious strategies, carefully thought-out maneuvers in a long-term campaign, but they look that way. Lions in the Serengeti use the same game plan, killing the cubs in a pride after chasing the resident male away from his harem.[335] And infanticide in similar circumstances has been recorded now for dozens of other species, including New World monkeys, apes and man. The similarity in tactics is not coincidental. It is the result of a complex calculation in which competing programs are run against each other in the course of evolution, and natural selection settles on the one which, on balance, produces the greatest genetic dividend.

Human social interactions are subject to greater input by casual, wilful, unruly interventions, but remain accessible none the less to mathematical analysis in the same way as lion and langur behavior. So much so that a whole new field has grown up around evolutionary genetics. The mathematics was originally designed to deal with economic prediction, but perhaps because economics is such a voodoo science filled with even more whimsical and unpredictable variables, it has found more ready application in biology, particularly in the study of evolution.[397]

At the heart of this whole new approach lies one of our oldest dilemmas. We know that people tend to look after themselves and their own, giving priority to relatives and friends—and we understand why that happens. We also have personal experience of the pleasures and benefits of cooperation. We appreciate how important generosity can be and, if asked, tend to agree that everything we value about our civilization is based on unselfish behavior. But where does it begin? Nothing in our experience gives good cause to expect cooperation to arise from situations where individuals have every incentive to be selfish. And yet it happens. Time and time again, we surprise ourselves and each other with acts of outright, inexplicable altruism.

The origin of such strange selfless behavior was a major mystery until Bill Hamilton showed in 1963 how it might evolve. Now, awareness grows that everything need not necessarily be based on the genetic imperative to "Look Out For Number One." That remains a major influence, but there are other forces involved, and to everyone's amazement these turn out to be ones which act in favor of cooperation. They suggest that even the most self-seeking systems can, without foresight or order imposed by an outside authority, gravitate towards genuine reciprocation. This discovery is so extraordinary that it is hard to take on trust in a world poised on the brink of epidemic anarchy. But, happily, there are hard numbers to back the new revelations—and the most convincing of these come, of all places, from computer games.

Game theory pictures members of a population playing games against each other and considers various permutations of behavior and the equilibria which can arise. It makes only one assumption, which is that self-interest will be in action in pursuit of as much fitness as possible, giving the genes involved the best chance of survival. Biologist John Maynard Smith calls this ideal solution an "evolutionary stable strategy," a look at what an individual ought to do in any situation in which it may find itself. In a perfect world, with every individual in a population using the strategy, no rival course of action, no mutant gene, could ever invade the system. Nothing of course is that perfect, it would be a very dull world indeed if it were; and no evolutionary system is ever in equilibrium for long. But game theory is producing some astonishing and counterintuitive results.[268]

Imagine a situation in which two teenage gang members are arrested and accused of a break-in, a crime which carries the penalty of a year in jail or a $100 fine. There are no witnesses, so the boys are held and interrogated separately, with the usual attempts being made to get one to incriminate the other in the hope of improving the police department's conviction record. If neither one does, if they keep to the code of silence, both must be set free and become eligible for the gang's usual bonus of $300 for loyalty. The police know this and try to crack the code and the case by tempting each boy to defect in return for a free pardon and a reward of $500. If both give in to the temptation, both get convicted, neither gets any reward or bonus and each gets fined $100. But if either boy is the only one to take the bait, the other gets the maximum sentence, going to prison for a year without the option of a fine.

These are the circumstances of the "Prisoner's Dilemma." It has four possible outcomes. If you think your friend can be trusted, you can keep quiet too, go free and pick up the $300 bonus. Or you can talk to the police and go free, taking with you the larger reward of $500. It pays you an additional $200 to defect. If, on the other hand, you think he is likely to talk, and you decide to do so too, each of you gets just the $100 fine. But if he talks and you don't, he goes free with $500 and you get the prison sentence. Once again, it pays you to defect. That is the logical thing to do.

The lawyer defending you knows the odds and advises you to talk. But the lawyer defending the other boy is equally astute and is likely to offer him the same advice, so the chances are that you are both going to suffer. In these circumstances, it is obviously best for both of you, individually, to cheat. To choose to enjoy the possible altruistic behavior of the other, without reciprocating. If you go for the big reward and are wrong, the worst that can happen is that each of you will end up with a $100 fine. But if you are right, and he chooses to ignore legal advice, it is you who goes free with $500 in your pocket. You have to cheat. Nothing else makes sense in such an encounter.

There is, however, a catch. Life is not limited to single encounters. It tends to be built up from long series of similar events, in which the outcome of each encounter remains unpredictable, but depends also on what happened last time. We have records that can be taken into account, and there are additional prices to be paid for past behavior. The courts can increase the punishment for breaking the law or, worse still, the gang can exact revenge for breaking their code of silence. So in the long run, over a number of encounters or moves, different strategies will have to come into play and these may encourage cooperating to begin with, and cheating near the end of the game, opting perhaps for the witness protection program only as a last resort.

There is, of course, always the option of reciprocal altruism, in which both participants choose to be trusting and to pay a recognized price for their mutual benefit. Both boys keep quiet, keep the police guessing, the gang happy and earn $300 each for their trouble. This seems to be the Panglossian solution, "the best of all possible worlds," but the problem is that it is difficult to see how it could evolve. How, in Darwinian terms, it could be justified in cost-benefit terms to begin with. And yet it happens, even to vampire bats.

Vampires live, not in Transylvania but in South and Central America, where they emerge at night to suck blood, tapping it with their razor-sharp teeth and grooved tongues from the exposed flesh of their prey, sometimes human, but more often these days domestic cattle or horses. It is a fine art; young bats have to learn the necessary stealth, but even adults quite often fail, getting brushed off by wary animals and forced to return to their nests unrequited. This can be a problem. Blood is good rich food, but so easily digested that vampires need to feed often. They literally crave blood and failure to find it can lead to starvation within three days.

This need puts considerable pressure on vampire bat individuals, but they seem to have found a sanguine solution by turning to the ideal of reciprocal altruism. The bats roost in communal groups which may or may not be closely related, and yet still share their meals. Bats who return well fed regurgitate part of their bloody supper to others in need. All individuals fail at some time or another, so the roles of blood donor and recipient are regularly reversed and confined to those bats who know each other well. These are key requirements for reciprocity, and are enforced in vampires by "moralistic" anger directed at any cheating bat who fails to reciprocate, who takes but doesn't return the favor of transfusion.[417]

It seems that two-way relationships of this kind, which depend on individual recognition or kinship, are essential for successful reciprocal altruism. They involve discretion and long-term memory and may well have played a major part, not just in vampire, but in human evolution. Biologist Robert Trivers suggests that in addition to learning how to cheat, it became necessary also to learn how to detect cheating in others, and to find elaborate ways of avoiding being thought to have cheated.

Humans share food, and we help the young, the sick, the poor and the wounded. And we do this routinely now, extending aid to those to whom we are not even distantly related. Such charity apparently has its roots in history, going back perhaps to a time when our lives were long enough, our groups small enough, our parental care extended enough, and our movements still restricted enough, to provide the sort of extended contact with closely related and clearly recognizable relatives that could give rise to altruism. We are talking about a million years ago or more, but Trivers suggests that it was then that we learned about "friendship"; a disposition to like others enough to want to enter into reciprocal relationships with them; "fairness": a sense which, coupled with moralistic aggression, guarded against cheating in reciprocal relationships; "gratitude": an appreciation of the cost-benefit ratio of reciprocal acts; "guilt": an emotion designed to repair relationships when cheating has been, or is about to be, discovered; and "justice": a sense which encourages common standards by means of which altruism could be recognized and assessed when more than two parties are involved.[386]

There is some evidence to show that group size can change the dynamics. If, for instance, a number of friends dine at a restaurant under an unspoken agreement to divide the cost equally, and one is tempted to order the expensive lobster instead of the modest dish of the day. He or she is more likely to be unscrupulous and selfish in a large group, where the additional cost is spread more widely and might even to unnoticed. The same rationale applies to social programs concerned with recycling or pollution. Individuals tend to behave more thoughtfully, more sensitive to the common good, in small communities than larger ones.[147]

All of which makes good sense, both logically and psychologically, but it is still difficult to imagine what exactly allowed reciprocal altruism to appear in the first place. Genetic Rule Number Two, which requires you to be nice to close relatives, is a good start; but what circumvented Rule Number Three? What was it that overcame the overwhelming advantage that cheating offers over honesty? There are no easy answers, but we do seem to have at least one line of evidence which suggests what might have happened, and this comes from recent extrapolations of the Prisoner's Dilemma.

Robert Axelrod is an American political scientist with a fascination for Machiavellian tactics and modern game theory. He entered the evolutionary argument with one simple question: "When should an individual cooperate, and when should an individual be selfish in an ongoing interaction with another person?" And he looked for guidance from an extended form of the Prisoner's Dilemma into a game which involves a long run of 200 choices, each of which has to be made on the basis only of knowledge provided by previous choices. In an inspired move, Axelrod decided to run a Computer Prisoner's Dilemma Tournament and invited experts in economics, psychology, sociology, politics, mathematics and biology to submit programs for the most effective long-term strategy—the mathematical equivalent of an evolutionary stable strategy. He chose sixty-two entrants as finalists, running their programs randomly against each other, and discovered to his surprise that the one which won was the simplest strategy of all.[21]

The program was called "Tit For Tat." It had only two rules. "On the first move, cooperate. And on each succeeding move do what your opponent did on the previous move." In practice what this means is a strategy in which you are never the first to defect; but in which you retaliate after your opponent has defected; and that you forgive that opponent after one act of retaliation. In short, it is an optimistic program which takes an altruistic act towards you as an open invitation to be altruistic back. It requires trust, which seems unrealistic, but the astonishing thing about Tit For Tat is that it not only invites reciprocity, but also provides the ideal mechanism for ensuring that you get it. Tit For Tat is a robust program which, once running, is almost impossible to stop. And because it is so simple in its instruction, it provides the perfect vehicle for genetic inheritance. Even bacteria are capable of direct responses to their environment and can certainly respond to orders which say, in effect: "First do unto others as you wish them to do unto you, but then do unto them as they have just done to you."[388]

Tit For Tat is altruism with teeth. A program that works because it carries a big stick. And it goes on working because it also forgives, recognizing and rewarding acts of remorse. All of which sounds impossibly high-principled, but the very real beauty and strength of the program is that it carries the seeds of its own success. It is self-realizing. An automatic process that leads inexorably to a distinct goal, creating good out of evil; just as selfish fish, each interested only in its own survival, accumulate inevitably into a school with very different characteristics.

Axelrod takes issue with those who argue that life in nature is necessarily "solitary, poor, nasty, brutish and short."[187] Thomas Hobbes used that prescription in the seventeenth century in favor of his view that cooperation could not develop without central authority in the shape of a strong government and an effective police force. There are still those who present such ideas as a valid political platform, and get voted into power on it. But what Tit For Tat shows is that we don't have to deal only with the three rules of genetic selfishness. There are strategies which, without genetic involvement, can allow cooperation to evolve. It is possible for individuals or species, each intent on pressing their own selfish interests, to gravitate towards reciprocity and mutual advantage. Axelrod points out that systems such as the United States Senate, despite being made up entirely of individuals all trying to look good at each other's expense, can still, through "folkways which involve helping out a colleague and getting repaid in kind," engage in enough mutually rewarding activities, such as vote-trading, to be seen as an organization that has become essentially cooperative.[20]

It looks as though cooperation can begin in small ways and that it can thrive on simple instructions that are essentially "nice," yet capable of being provoked, while also being somewhat forgiving. And that systems which start in this way tend to grow; they contain a "ratchet" which ensures that levels of cooperation remain the same or increase, but never go down. Axelrod, the political scientist, points once again to the United States government as an example, this time to the Congress, noting that: "In the early days of the republic, members were known for their deceit and treachery. They were quite unscrupulous and frequently lied to each other. Yet, over the years, cooperative patterns of behavior emerged and proved stable." He may be too kind in this case, but the point is well taken. Cooperation flourishes best in *continuing* relationships where the participants can anticipate mutually rewarding transactions in the future. And such anticipation is by no means exclusively human. All that may be necessary is a basic awareness of a simple and sobering social reality. "We may meet again."

In 1976 I was sailing along the shore of the Dry Tortugas Islands west of the Florida Keys when I noticed a disturbance in the sandy shallows that looked like a dark, writhing many-headed monster. There was an intermittent haze over the hydra, but the sea was calm and I was able to leave the channel and get close enough to see that the creature was in fact a pod of small whales, lying in just enough water to support their sleek black bodies. And by the end of that day I found myself part of a small fleet of boats, whose crews all joined Coast Guard and National Park Service personnel in helping to avert what ranks still as one of the best documented and most revealing of all whale strandings.

They were false killer whales, slender sixteen-foot deep-water predators of fish and squid, which are known from time to time to come ashore in close-knit groups. There are many theories about such strandings, ranging from navigational errors to panic and self-destructive tendencies. But it was clear on this occasion that the problem lay with a large male, probably the oldest in the pod, who was bleeding from one ear and had so much trouble remaining upright that he would have drowned had he not been lying on soft sand supported by the bodies of the rest of the group, who pressed in close enough to keep him steady with his blowhole above the water.

We tried at first to lead some of the twenty-nine apparently healthy whales on the fringe of this "clinic" to deeper water, but they became highly agitated and we discovered that no amount of human effort could prevent two tons of whale from returning directly to wedge itself into the support system back in the shallows. There, as soon as they were in direct contact once again, the high-pitched alarm subsided into a lower frequency conversation that carried on continuously day and night. We stayed with them through three whole days, listening to what was being said, trying to understand, doing what we could to keep their backs wet and protected from sunburn during the ebb tides; but none of us, whale or human, could do anything more until the male at the heart of the happening died on the third night. By dawn, the tight group had relaxed and disbanded, the constant chatter and squeal of reassurance was muted and, as soon as the tide was high enough, the survivors left of their own accord—clearly aware, as we were, that the ailing male was now beyond concern.[407]

I use the word "concern" here with good reason, believing it to be an accurate description of the behavior of the whales, because each time one of us swam anywhere near the group wearing a snorkel—through which we make noises that do sometimes resemble those of a whale with a waterlogged blowhole—we would be promptly "rescued." One of the whales would detach itself from the group, slide underneath the human swimmer, rise so slowly that he was lifted almost clear of the water, and carry him further into the shallows. It was a humbling experience to be so propelled, able to do little to interfere with the effort until one could stand and take out the mouthpiece and stop making such distressed and distressing sounds.[306]

It seemed clear, throughout the seventy-odd hours that we spent together, that the whales were involved in an act of altruism towards one of their number who was going to drown unless he was assisted to breathe. It was obvious that any of the other whales could have left him at any time, and did not do so even at risk of their own lives. And it became apparent that the altruism was real and reciprocal when the whales we were trying to help came just as readily to our aid when we too seemed to be having trouble with our breathing. Helping across the species lines argues very strongly against a simple genetic explanation for altruism. I am not even remote kin to a false killer whale, but I have absolutely no hesitation in saying that, as a direct result of that experience, I am now so kindly disposed to that species that I would do anything in my power to cooperate with any of its members if we ever meet again. When faced with altruism, it is very hard *not* to reciprocate and to play good Tit For Tat. It is a pattern of behavior with definite survival value.

It may be in the nature of life to be selfish. That which survives, survives—and goes on to fill the world with its own ruthless qualities. That is how Darwinian evolution works, building up ever more complex and more self-serving societies. But there is a ratchet in the works, a mechanism which allows something more generous to grow out of indifference, something "good" rather than "bad," something working towards the Goldilocks solution. Something very like the live-and-let-live reciprocity that arose between opposing troops on the Western Front during the First World War.

Trench warfare put enemies into very close contact with one another for extended periods of time, providing one of the vital prerequisites for cooperation, the anticipation of meeting again. Ad hoc truces, periods of mutual restraint, sprang up all along the lines, resulting in times of quiet honored by both sides, during which troops could take care of their wounded, bring in rations and even bury their dead without fear of falling to a sniper's bullet. Headquarters hated it, of course, issuing orders making clear that the soldiers were there to fight, not fraternize with the enemy. The high commands of each army were afraid that the tacit truces would sap the morale of their troops, who were becoming more and more expert at making peace than waging war, and squashed the growing tendency by introducing a new policy of incessant aggression. They had their way, but not before it became obvious that cooperation can develop even between sworn antagonists. And it was informal contacts and casual covenants of this kind that contributed at the end of that war to a mood that made establishment of the first League of Nations possible.

Game theory has proved useful to evolutionary studies because it pictures members of a population or a species playing games against each other in a Darwinian struggle for survival. But recent awareness of the value and inevitability of altruism makes it necessary to look for a different kind of game. One perhaps in which the rules concentrate on cooperation rather than rivalry. As it happens, there is such a game. It is called *kemari* and was once popular with the Japanese imperial family. It involved a sort of soft football that was kicked around a circle of players, with each participant concentrating all their energies on not being the one to let the ball drop. There are no winners, only losers whose penalty is to leave the game.[118]

In life, that means forever. Extinction seems to be a fitting forfeit for those unwilling or unable to keep the ball in play. And survival with the advantage of consciousness—knowing that you are alive—a proper reward for rising above the dictates of the gene.

It is clear that genes are selfish; but that organisms need not be. We may be cowardly creatures, born selfish and subject to the ruthless interests of genes that encourage cheating and lying; but we are also part of kinship systems that put a premium on being nice to close relatives. Genes are primarily concerned with inclusive fitness, with the big picture; but parental and social behavior require a certain amount of selflessness which gives us, as individuals, the experience of generosity and sympathy. There are distinct advantages to be gained by deception; but the very existence of such wiles produces the need for awareness of them that has made us calculating and intelligent beings with a sense of justice and fair play. Self-interest may be guaranteed; but kinship, individual recognition and extended contact all provide the conditions necessary for altruism to appear in contradiction to the three rules of the genes, and once it does, there are simple mathematic principles, rules of the universe, in play to ensure that they increase and encourage cooperation instead of conflict.

It begins to look as though there is something in "original sin." There is an inherited, genetically related system that is unrelentingly selfish, ruthless and cruel. And Saint Augustine is right, we are never going to be without it. None of us is innocent, not even as a child. We are congenitally "bad," along with the rest of nature. It is important that we should be, because natural selection makes no moral judgments. It recognizes only success, and measures this only by possession of those qualities most necessary to survive. More often than not, these are hard ones, inevitably ignoble. In the words of Richard Dawkins, in penitent mood for his preface to Robert Axelrod's book on cooperation: "As Darwinians we start pessimistically by assuming deep selfishness, pitiless indifference to suffering, ruthless heed to individual success. And yet, from such warped beginnings, something that is in effect, if not necessarily in intention, close to amicable brotherhood and sisterhood can come."[21]

Note the caveat "not necessarily in intention." That is an important qualification. There is no evidence, in the transition from brutish nature to altruistic human concern, of redemption. No sign of a conscious change of direction brought about by the imposition of a new plan involving the cancellation of past debts and a brand-new beginning. If we are sometimes "good," it is purely and simply because we are also "bad." If we are nicer than hyenas, we have them to thank, because they and their kind are part of our history and our complex inheritance.

Fortified by the biological scene-setting of these opening chapters, knowing a little of how the world and we came to be so precariously perched, so finely tuned, I think it is time we left all the ecology and arithmetic behind—and looked human frailty full in its yellow eye.

Is "evil" just the sum of what we should expect from a species with our background? Is it human nature to take the dark side of genetic selfishness to unpleasant excess? Or is something else going on? Despite a trend towards cooperation, could there be "something wicked" in the air? Something which happens to be not "just right," but deliberately, destructively "wrong"?

Something with its own strange agenda . . .

PART TWO

HUMAN NATURE

THERE IS A CAPACITY OF VIRTUE IN US, AND THERE IS A CAPACITY OF VICE TO MAKE YOUR BLOOD CREEP.

Ralph Waldo Emerson

The son of Harmen Gerritszoon painted portraits, roughly ten a year throughout his working life. These display not just complete mastery of light, shadow, color and technique, but a profound knowledge of facial expression, gesture and posture. He was able to get under the skin of his sitters in a way few photographers can, producing images that captured the essence and realized the humanity of those portrayed, providing greater psychological insight than any other painter has ever achieved. It will come as no surprise to learn that the given name of this miller's son was Rembrandt, but it is interesting to learn that the paintings also contain more information than even he realized or intended.

Portrait painters seldom show their subjects full-face. They tend to angle the head to one side, providing a more three-dimensional, more characterful, image. There is no mechanical or aesthetic advantage in turning the head left or right, no evidence that the circumstances of a sitting favor any particular bias, and no tradition that dictates which is a sitter's "best side." But Rembrandt's choice in the matter was anything but random.

Psychologist Nicholas Humphrey analysed 335 Rembrandt portraits, fifty-seven of which portray the artist himself. Of these just nine give prominence to Rembrandt's left side; in all the rest—over eighty percent—he can be seen turning the other cheek. We all have idiosyncratic images of ourselves, and it may be that he was self-conscious about a bump over his left eyebrow, but the bias of the other 278 portraits suggests a different conclusion. Some of these, another fifty-six paintings, involve his immediate family, and if these are divided into those who are blood kin, and those who are relatives only by marriage or common law, a fascinating split becomes apparent. Over seventy percent of the portraits of Rembrandt's father, mother, sister, brother and son show *their* right cheeks. While less than thirty percent of the paintings of his wife and mistresses do so. And just twenty percent of the portraits he painted of other women, to whom he was not related at all, are "righteous."[201]

No moral judgements are made or implied in this, but it looks as though Rembrandt, perhaps unconsciously, allowed his portraits to reflect, among other things, the social distance between himself and his subject. There is a direct relationship between the numbers of sitters facing left, showing their right cheeks as he preferred to do in self-portraits, and his genetic connection to them. And the great majority of family members angled in this way are his closest kin, those with whom he shared fifty percent of his genes. Wives and mistresses, with no genetic link, nearly always look the other way.

There are no known portraits of cousins, grandparents or grandchildren—those with a twenty-five percent share in the Rembrandt gene pool—but if such are found to exist, then it becomes possible to predict that they will fall roughly halfway between family and friends, between genetic insiders and outsiders. Half of them are likely to face each way.

Nick Humphrey, never slow to look for larger implications in such things, suggests that we have in Rembrandt's paintings the material for a kind of "paleo-psychology." He points out that in the celebrated group portrait of *The Anatomy Lesson of Dr. Tulp*, now in The Hague, the surgeon shows his left cheek while the students, without exception, display their right. Perhaps Rembrandt, with his humble background, unconsciously identified with the students, as those like himself who were hungry to learn about the world, rather than born to teach others how to do so.

Rembrandt's portraits are certainly coded in a way that shows a very refined sense of relationship. This may be no more than was usual for a sensitive person of his social standing in seventeenth-century Holland. It may also be something much more basic, altogether more profound. Evidence of a fine biological sense we all have of each other, a natural feeling for the niceties of association which could be particularly useful to a portrait painter. Perhaps all painted portraits, the best of which are the product of a bargain between artist and sitter, have such potency. They provide depth and insight because they are true "likenesses," capturing something usually available only to close association. Speaking volumes about character, social standing and genetic affinity—and doing a lot of this with no more than a gentle tilt of the head.

Christopher McManus joined with Humphrey for a broader survey, examining 1,474 portraits painted in Europe between the fourteenth and the twentieth centuries. It is not possible to make a genetic analysis of these, but a clear majority—over sixty percent—showed the sitters left cheek on, suggesting that there was little or no social or genetic connection between the artists and most of their subjects. There was however a sex difference, with more women sitters showing their left cheeks to the painters who, being male, may have felt socially closer to their male subjects. Individual artists will inevitably have had their own feelings about personal and social distance, but the fact that their collected work produced the same 60:40 ratio for left versus right cheek exposure as Rembrandt's portrait *opera,* suggests that the bias is more biological than logical, more intuitive than intellectual.[253]

Either way, it matters. In 1885, while working on his powerful painting of two couples eating in a simple country home, Vincent van Gogh wrote to tell his younger brother of a lithograph he had made from a preliminary sketch for *The Potato-Eaters.* He copied this sketch directly onto a stone block where it produced a mirror-image print, but this so upset him that he told Theo: "If I make a picture of the sketch, I shall make at the same time a new lithograph of it, and in such a way that the figures which, I'm sorry to say, are now turned the wrong way, come right again."[395]

It is clear that there is nothing neutral about the composition of this painting, now in the National Museum in Amsterdam. It doesn't work so well, the message is confused, if it is reversed so that the two men apparently reveal their left cheeks and the two women their right cheeks, instead of vice versa. Tests have shown that even ordinary people, with no training in art, can pick out the mirror image of an unfamiliar picture from the original, provided that both perspectives are shown to them at the same time. There is something "wrong" about the mirror image, often very wrong if it happens to be a good portrait. Van Gogh, it seems, operated under the same constraints, using the same sensitivity and the same sign-system as Rembrandt before him—forced to paint—himself "right" side out, even after slicing off the ear on that side in 1888.

Few of us have their talent, but perhaps all of us share their instincts about those around us. We pick up a lot more than we realize. One of our special talents, perhaps the most important one of all, is our skill at reading one another and of making value judgments about each other that influence our association.

To the casual glance, humans seem to be bilaterally symmetrical, like most other free-moving animals. The environment favors such shapes, but appears to be neutral about laterality, providing no natural pressures that favor right over left or make us more responsive to one side or the other. Legs and wings are symmetrically arranged in nicely even numbers on each side of our bodies so that running, swimming or flying carries us along a chosen straight line, rather than round and round in obligatory circles. Sense organs also tend to be balanced to make us equally alert to information coming from either side. All of which makes sense and suggests that if life exists on another planet, it will inevitably find, even if conditions are very different, that it is shaped by the same natural needs. But look more closely and you find that this outward appearance of symmetry is all an illusion, often a deliberate "lie."

Symmetry is not a natural property of life. Our fundamental molecules, particularly the amino acids that form the building blocks of proteins, are asymmetric. They exist in two mirror-image forms, but nature only uses one of them. The genetic threads of DNA could corkscrew to the left or right, but in practice they always are right-handed. It is clear that nature has gone out of its way to produce organisms that are largely symmetrical out of components that are patently asymmetrical. Such pains served the purpose of giving us the mechanical and behavioral advantages of bilateral symmetry, making us suitably streamlined for life on the run, but there is another and even higher purpose served by keeping things also a little out of true.

We are lopsided animals. Not so skewed as the flounder, lying on its side, whose one eye has been forced to migrate to the other side of its face rather than stare forever at the darkness of the ocean floor. But asymmetric nevertheless, with right ears on most of our heads placed lower than the left, and one eye larger than the other. The right testicle in most men is larger than the left and not quite so pendant, while the left breast in most women is larger and lower than the right. Internally, there is not even a pretence at symmetry. In the interests of efficient packaging, our hearts and spleens are on the left, our livers on the right. Perfect symmetry is not only impossible, but also unnatural and uninteresting. Portrait painters know this and make the most of every difference, but their statistical preference for doing it from the left remains mysterious.

A clue to this imbalance lies perhaps in the fact that we are not just physically, but behaviorally and psychologically asymmetric. Every time you wave or wink, clap your hands or shake your fist, fold your arms or cross your legs, you betray a bias. And this begins very early. At three months, a human baby uses both sides of its body with equal vigor. At four months, it shows a preference for the left, which shifts at the age of seven months to the right. By the ninth month, most lean left again and return to this proclivity in their eleventh month, while showing rightist tendencies in their tenth and twelfth lunar periods. This vacillation continues until relative stability takes over in about the fourth year of life and is finally and permanently fixed by the age of eight, when nine out of every ten people become dominantly right-handed, eight out of ten right-footed, seven out of ten right-eyed, and six out of ten right-eared.[305]

These are global figures, independent of race, sex or cultural background. Whatever our superficial symmetry might suggest, we are a right-minded species. Not in the sense of using one cerebral hemisphere more effectively than the other, but in the fact that we show a consistent and predictable bias towards the right in almost everything.

Right is "right," right? And left is, by definition and default, obviously "wrong." To be a *right-hand man* is to be normal, important and useful. To be a *left-hand wife* is to be a mistress. A *left-handed compliment* is an insult; a *left-handed diagnosis* incorrect; and a child born *on the left side* or *bent to the left* is either illegitimate or homosexual. I once owned a boat which the previous owner called *Lefty* and sold to me far too cheaply. It sank before I got the chance to use it. Buying it was a *left-handed deal,* something best avoided.

Such prejudice is not peculiar to English. It seems to be universal. The French word for left is *gauche,* which can also mean "ugly" or "clumsy." Italian *mancino,* a left-hander, is derived from the root word used to describe anything "maimed" or "deceitful." The German *linkisch* can be defined as "awkward" or "maladroit." The Spanish *zurdo* comes from *azurdos,* meaning "to go the wrong way." And in Gypsy Romany *bongo* means not just left-handed, but also describes a crooked card game, a fixed horse race, or a wicked person. And the root of all this insulting confusion lies probably in the Latin word for left, which is *sinister*—and that, quite unequivocally, means "evil."[82]

Experts differ on the causes of lateral bias, most tending to look on the minority condition of left-sidedness as abnormal and in need of explanation. The fact that it is less common in older people supports one theory that cultural pressure makes us right-handed, such pressure being stronger in the days when our old were young. But this is contradicted by some studies which show that left-handers do tend to die younger than right-handers. A suspicion that sidedness may be inherited is widely held, but has also failed so far to find convincing genetic support. Nothing concrete has come from attempts to detect consistent neurological damage, or differences in the structure or function of the brain, in left-handed humans. And finally, arguments from evolution have been stymied by the discovery that, if there is any observable bias in any other species at all, it seems to be towards the left, not the right.[80]

The origin of handedness remains mysterious, but its reality is beyond question. Ten percent of the human population differs from the rest in dramatic ways, and this fact alone has produced inevitably negative reactions. The left, everywhere, is regarded as unlucky. If you spill salt, throw a pinch over your left shoulder, because that is where devils tend to lurk. Step into your home with the right or lucky foot first, and make sure that your guests do the same. The Romans even posted a slave on the threshold of each villa whose sole task was to enforce the custom. He was, presumably, the first "footman." Blessings and benedictions in most traditions are given with the right hand, while unclean and unseemly things are reserved for the left. The Judeo-Christian devil himself, of course, is left-handed, leaving matching talon marks on the bodies of his acolytes. Those suspected of witchcraft were always stripped naked in search of such blemishes, and were sometimes convicted on the basis of nothing more than a mole on the sinister, left side of their bodies. Suspicion, naturally, fell most swiftly on the ones who were already clearly left-sided.[29]

The French humanist Montaigne condemned such peremptory intolerance, saying: "It is rating one's conjectures at a very high price to roast a man alive on the strength of them."[275] But prejudice operates on such a basis, finding easy reasons to punish other people, quickly choosing to put them outside concern if there is any doubt about their affinity. "Different is bad," is just another way of phrasing the first genetic rule: "Be nasty to outsiders." Things work that way in nature. The freak or "sport" is chased out of the herd *because* it is different; and because, if it is different enough, the chances are that it has mutated and represents a genuine threat to the integrity of the resident gene pool. It is eliminated in the same way as an animal breeder concerned with purity singles out and selects against individual cats with kinks in their tails or dogs with unfriendly dispositions.

It has been suggested by those who believe in a genetic cause for left-handedness that such genes are linked as well with an increased tendency to be accident-prone, schizophrenic, depressed, suicidal and criminal—but all these conditions are just as likely to be the simple and direct result of being left-handed in a world designed entirely for the convenience of right-handed people. Scissors, can openers, coffeemakers, bread knives, food slicers, drill presses, band saws, rulers, spiral-bound notebooks, computer keyboards, fishing rods, steering wheels and even shirt buttons are frustrating, often painful and sometimes dangerous in the hands of ten percent of the population. A minority who also, perhaps as a consequence of having to deal with such discrimination, have produced more than their fair share of superachievers of a wayward kind, such as Leonardo da Vinci, Charlie Chaplin, Julius Caesar, Babe Ruth, two of the four Beatles and an astonishing four out of the last five Presidents of the United States.

It makes sense somehow that the aloof George Bush and the accident-prone Gerald Ford should have been left-handed, and even the long-haired young Bill Clinton fits part of the rebellious sinistral stereotype, but would you have guessed that Ronald Reagan leaned towards the left? Probably not. And yet Stanley Coren, probably the leading authority on the causes and consequences of being left-handed, recalls a conversation with a White House aide which suggests that Reagan was indeed naturally left-handed, though he seems to have gone to extraordinary lengths to conceal that fact—writing, throwing and gesturing with his right hand at all times in public. As part, perhaps, of a conscious campaign to be seen as an "insider," one of us, someone in and of the "right," instead of a man of only ten percent of the people?[82]

We have strong convictions about what makes people "good," what constitutes the "right stuff." And we reserve our special acclaim for those amongst us who seem to have the most of what it takes. We call them the "beautiful people," the few, and assume that true beauty is exceptional. Right? No, once again, probably not.

Evolution is actually very democratic. It brings pressure to bear on the extremes of any population, operating against those with too much or too little of what we consider proper. Those with average qualities turn out in the end to have the best chance of survival. And inclusive fitness is best served by being attracted to, and mating with, someone who comes as close as possible to the average value.

Proof of this tendency came first from a pioneering study made in 1878 by the British anthropologist Francis Galton. He used a stereoscope to superimpose full-face photographs precisely on top of each other to produce what he called composite portraits or generalized pictures. He found, to his surprise, that these were by no means confused. They portrayed no one in particular, but had a distinct air of reality. "Nobody who glanced at one of them for the first time, would doubt its being the likeness of a living person."[137]

Galton worked with whole families, producing composites which come as close as anything can to a genetic portrait. He compiled sets of criminals convicted of murder or violent robbery, and found that the process smoothed out their obviously rough edges, making "the features of the composites much better looking than those of the components." And he quotes from a study referred to him by Charles Darwin in which a similar technique using portraits of fashionable ladies produced "in every instance a decided improvement in beauty."[19]

The most conspicuous feature of the composites is their symmetry, an eerie perfection that, while bland, is somehow attractive. A vestige perhaps of a time when asymmetry was more indicative of injury or disease and warned of possible deficiencies in evolutionary genetic fitness. We have become so used to hearing that "beauty is in the eye of the beholder," and to assuming, because standards of beauty seem to be culturally specific and to vary so enormously, that any attempt to find universal standards would be futile.[203] Perhaps not. At the University of Texas at Austin, over 500 photographs of the faces of predominantly Caucasian students were divided into male and female pools, digitized by scanning, and mathematically and randomly blended by computer into composites of four, eight, sixteen or thirty-two faces. When these were analyzed for "attractiveness," the composites were consistently rated far higher than individuals of either sex, with the highest scores of all going to the most complex blends.[237] The more average the picture, the more individual portraits contributing to its creation, the more attractive it seems to be. And in follow-up studies in Scotland and Japan it was found that students there showed precisely the same preference, even when faced with blends taken from another race or culture.[299]

In addition, studies involving children, some only two months old, all show that even babies agree. When presented with pairs of photographs, they consistently looked longer at those rated "more attractive" by adults, showing the greatest interest of all in computer-generated composites.[237] It seems clear that beauty is not necessarily local nor limited to any race, place or period. Cosmetic companies, who sponsored some of the research, were delighted to learn that there is global agreement on what is beautiful and, where the human face is concerned, a distinct and universal preference for the ordinary rather than the extraordinary. True beauty, it seems, is just average.

We appear to be programmed, at least as far as mate selection is concerned, to prefer the ordinary, the unthreatening norm. But anyone who has ever been stood up in the dating game, abandoned in favor of a more exotic and dangerous rival—the leather-clad rebel on his motorbike, or the seductress in her tight silk trousers—knows that we also find the "wrong" things, the "bad" people, attractive. Perhaps because, like the male peacock with his outrageous tail, they send a new and provocative message of enhanced fitness, which says: "I can afford to be this flamboyant because, unlike you, I have the resources to back it up." The danger, of course, is that they or their genes could be lying. They often are.

Duplicity, however, is not limited to relationships with others. We who deceive also have the capacity to deceive ourselves. It is important that the truth is registered somewhere, but it can be equally vital that we carry off a bluff successfully from time to time, and to do that really well it becomes necessary to hide the truth from our own consciousness, running two sometimes competitive descriptions of reality at the same time. Under social pressure, we frequently find ourselves "in two minds" about something. To facilitate this dissociation, we have evolved two distinct and very different mind-areas, and this may be the most important asymmetry of all.

It has become fashionable since the 1960s to think and talk about two more or less independent brains within each human skull. These are usually described as the left and right hemispheres. The left, it is said, counts, marks time, analyzes, thinks logically, and governs spoken and written language. This is the dispassionate, calculating, masculine mind that controls the right hand. On the other hand, controlling our left side, is the right hemisphere which dreams, fantasizes, invents, perceives intuitively, and is responsible for art and music. This is the creative, emotional, feminine mind. And the idea of dichotomy between the two sits so well with us, feels rightly ambivalent, and corresponds so nicely with eastern ideas about a world governed by opposites such as *yin* and *yang*, that it has fuelled an extraordinary outpouring of work in support of the two-mind idea, and a new-age need to feed our underdeveloped "right brains."

The truth about the hemispheres is that they are connected by 200 million nerve fibers in the corpus callosum and that there is a great deal of overlap in the abilities and functions of the two sides. There is no good evidence to show that left-handers have their language centers in the right hemisphere or think less logically than right-handed individuals. And every reason to assume, from work with people who have lost the use of a whole hemisphere, that each side is fully capable of taking over all the functions normally carried out by both.

I am not suggesting that we have no mental dichotomy at all. I believe the evidence to be very strong in support of the idea that we do. But given the difficulty of locating the mind, I find it more useful to think of the asymmetry being functional rather than spatial, one of complexity rather than involving a specific location for every human talent. I am beginning to find those maps of the hemispheres as uncomfortable and anachronistic as the bald china heads with their confident patchwork of virtues which once graced Victorian parlors.

New Zealand psychologist Michael Corballis has, I think, come up with the most productive suggestion. In his description of humans as "lopsided apes," he argues that with the arrival of a bipedal, tool-making, language-using way of life—sparked largely by our right-sided laterality and the influence this has had on "left brain" complexity—we have had a qualitative change of mind. The structure of our brains remains essentially anthropoid, but a discontinuity has appeared, marked by a new emergent quality he calls "generativity," meaning the capacity to assemble, associate and represent all the old information in an entirely new way, giving direct rise to mathematics, reason, music and art. But, if we accept this possibility, then we must also accept the inevitability of products of another kind. The fruits of industry, greed, pollution and war. The two streams spring from the same source.[81]

What this means for our study of the origin of evil is that it is possible to point to a moment, a turning point in evolution at least as consequential as the emergence of flight in birds. One which, it could be argued, has led to human flight as well. Birds were preadapted to flying by the evolution of feathers for thermo-regulation, just as we were prepared for full human consciousness by the evolution of grasping hands for more effective feeding; but neither feathers nor fingers had any such achievements in mind.

I add to the idea of generativity the suggestion that, coming from lopsidedness, we have long been very much aware of asymmetry in ourselves and each other. And we have become skilled, as Rembrandt's portraits show, in recognizing and giving meaning to such inherent bias. We use it as a marker of what is right and wrong; of what works and what does not; what is good for life, and contributes directly to social or ecological equilibrium, and what is bad for the system and pushes things over the top into chaos.

This is a vital talent, the acquisition of which is rightly recognized in Christian myth as a turning point in human evolution—the moment of awakening, of the growth of knowledge of good and evil. It comes, of course, with another crucial awareness. "The eyes of them both were opened and they knew that they were naked; and they sewed fig leaves together, and made themselves girdles."[190] The first fruit of this awareness, we are told, was shame—which is an astonishing response. Being naked apes actually isn't easy. It leaves us exposed, vulnerable to the elements and our enemies, but also to each other—hence the fig leaves. Token clothes, intended not for thermo-regulation or protection, but for boosting self-esteem. No animal, as far as we can tell, feels shame or the need to hide from itself in this way. However it happened, such feelings are now part of our nature. We need to "look good" and we feel bad about not doing so. Nothing makes us more angry than being made to "look bad." Looking bad means being seen to be different, perhaps even deviant or maladept; in biological terms, badly adapted. And that puts us to shame, bringing personal or evolutionary shortcoming to conscious attention, making us aware of what is wrong, making us anxious to put it right, to conform. Shame is therefore a social sense, one that serves the community well, and having it, being concerned with convention, suggests that we may be essentially moral beings. The first ethical animals.[222]

I hope so. But before rushing to congratulate ourselves, we need to look at all the things we continue to do that are truly shameful, that go against the social and ecological norm and give us, in addition to our virtues, such an extraordinary capacity for vice.

THREE

BEING BAD: THE ETHOLOGY OF EVIL

As a biologist, I see no reason to exclude humans from my study of nature. We are animals and, until a truly impartial observer—some extraterrestrial Darwin in his warp-driven *Beagle*—publishes a definitive study of human natural history, we are going to have to compile our own field notes.

This is what biologists do. We look and listen, we make notes and report back. And with the help of others so occupied, we try to compile the biological equivalent of a composite portrait. We look for common patterns, for behaviors and attitudes which are typical to a species and tell us something about its inclusive fitness. We want to know what it is that keeps it healthy and what makes it successful. We try, in short, to put it into biological perspective and to understand its place on the planet. And if we have any experience in the field, we know that our study will have to begin with the usual basic questions. How does the animal in question behave? How does it solve the universal problems of feeding and fighting, of mating and rearing its young?

The most cursory glance at the human animal, in all its habitats and guises, warrants one quick and easy conclusion. The subject which most occupies the big brain of the naked ape, filling more of its mind and time, flowing over into almost all of its other endeavors, is sex. Freud was right about that. Sex has been a central concern in human evolution. Our sense of beauty, our capacity for love, our widespread tendency to be jealous and aggressive, even our vaunted intelligence—all the things which seem to make our species so special are natural consequences of our preoccupation with sex. And by-products of such obsession have been a sense of shame, an awareness of sin, and widespread adoption of the alibi of evil. "It wasn't me, the devil made me do it."

Why sex? The answer is easy enough. Humans exist because of the changes made in living matter through the long course of organic evolution. And evolution seems to work, almost exclusively at its more complex levels, by competitive reproduction. This is what gives natural selection its raw material, setting life on a new path with every generation, mixing and matching as it goes along, creating new strains with new advantages, passing along those that work best, favoring these at the expense of less effective strategies, doing everything possible to reproduce again and again. Everything else is simply a means to that end. And because reproduction in humans, with very few and possibly apocryphal exceptions, means sexual rather than asexual reproduction, we have put a premium on sexuality, making it the forbidden fruit. The first knowledge we gained from that tree in the garden was carnal knowledge, and this—more than anything else—has made us what we are.

What we are is human. And professional students of humanity, anthropologists, make much of the differences between the customs and rituals of various races and tribes. But what strikes a biologist most forcibly is how much alike we all are, how closely we conform to patterns typical of our species. Zoologist Matt Ridley, in his comprehensive and elegant summary of the role of sex in human evolution, puts his finger right on this crucial pulse with the observation that: "Human culture could be very much more varied and surprising than it is. Our closest relatives, the chimpanzees, live in promiscuous societies in which females seek as many sexual partners as possible and a male will kill the infants of strange females with whom he has not mated. There is no human society that remotely resembles this particular pattern."[321]

Why not? There easily could be. Our intellect and versatility could probably find ways of justifying it. We can talk ourselves into almost anything, and are particularly fond of solutions for which there is historic precedent. But the fact is that we are also strangely constrained. Part of this is purely anatomical. We are ill-equipped for rampant promiscuity, having testicles far smaller than most chimpanzees.

Polygyny, a social arrangement in which dominant males gather more than one breeding female, is the usual solution to reproduction in most of the animal kingdom. It remains the norm for three-quarters of the human species, but it has had to be modified for our use because of the relatively small difference in size between our sexes. A 600-pound silver-backed male gorilla can hold up to a dozen 150-pound females in respectful thrall. But a 150-pound human male has to compensate for his lack of physical presence by other subterfuges.

A lucky few build harems based on wealth and position. Atahualpa, Sun King of the Incas, is said to have retained 1,500 women at his pleasure in each of several houses set up at strategic points in his kingdom. Any man who challenged his hegemony and exclusive access to these women was punished by being put to death, along with all his family, friends, fellow villagers and llamas. Such excess, which condemned most other Inca men to celibacy, was obviously highly unstable.

The usual human solution has been a more restrained one. One sanctioned by tradition and tribal custom and given sanctity by law. *Polygamy* is our answer, an arrangement in which some males still gather more than one breeding female, but give each one formal protection by a contract of marriage. Such agreements have come to be questioned, but continue to thrive wherever each wife of a rich and successful man is going to be better off, more likely to reproduce well, than if she were the single wife of a poor and unsuccessful man. Which woman, it has been suggested, would not rather have been John Kennedy's second wife than Lee Harvey Oswald's first?

In Africa, polygamy is so common still that many states even recognize a first wife's primacy, giving her legal right to seventy percent of her late husband's wealth. But in most of the first world, where it is now illegal to have more than one wife, such statutes arrived on the books not so much to protect women's rights as to protect men, the majority of men, who might otherwise be unable to find a mate themselves. The institution itself is not unbiological and not even necessarily unattractive, even in the twentieth century. In Mormon Utah, where polygamy is still practiced by complicity, without automatic interference from the law, there are said to be compelling social factors—not least of which is the possibility of reliable home help with child care—that make plural marriage "very attractive indeed to the modern career woman."[157]

One of our close relatives has gone to the other extreme. Gibbon groups are confined to two adults, a pair practicing monogamous fidelity in isolation from others of their kind, keeping contact with, but at the same time maintaining distance from, the rest of their species by elaborate defensive songs or duets. "We are the Hoolocks, and this is where we live, but we would rather you didn't drop by."

We could have made the same choice, but we didn't, and this time the constraints were largely social and psychological. Our peculiar history does invite pair-bonding and territoriality, but we were also once pack hunters, vegetarians turned carnivore, foragers who shared food and mounted a collective defense, and this cooperative experience, which has filled ninety-eight percent of our human history, has left its marks on our minds and in our genes. Our human culture continues to be directly affected by our animal origins. We value our privacy, but also share our private spaces with family and friends. We have a natural tendency to be gregarious, even outside the home—and the important part this plays in our lives, the degree to which we still take every opportunity to join groups, clubs, societies and other associations, shows how new and artificial we find the advertiser's dream of the nuclear family in a "single-family home."

The human mating system is complex. As individuals, and more commonly in the last few generations, we tend to do more or less what we want. But collectively, there is still a pattern. There is a disposition towards monogamy and long-term bonding, which leans on a tendency we have to pair, even if only "for the sake of the children." There is also, however, a tendency, particularly among males, to stray beyond the confines of the bond, taking additional wives where this is allowed or can be afforded; and to indulge in adultery as frequently as custom and circumstances will permit.

Such behavior obviously can be disruptive, but it has very deep biological roots. Sex, like life itself, is asymmetric. There is a basic inequality involved in it which begins with the relative size and value of the sperm and the egg. A man produces several hundred million sperm a day. A woman just one egg a month, perhaps 400 in a lifetime, fewer gametes than her mate makes every second. A sperm is just one five-hundredth of an inch long, and most of this is tail. The business part, the head which contains twenty-three chromosomes with 75,000 genes, is an order of magnitude smaller, perhaps one five-thousandth of an inch in diameter. The volume of the egg is 85,000 times greater, holding its twenty-three chromosomes in greater safety, and with astonishing patience. A woman's lifetime supply of eggs are poised, stacked up in the ovaries from the moment of her birth, ready and waiting for their turn to travel to a rendezvous in the Fallopian tube.

In the economic language of genetic fitness, women may be described as living on capital. Sexually speaking, they have no income. While men are entrepreneurs, starting each new enterprise with nothing, producing what they need, taking risks or taking out loans, as necessary. Women spend an average of 266 days internally nurturing babies that are enormous and highly invasive compared with the offspring of most other primates. Men become biological fathers in as many seconds. Men can increase their fitness by having additional mates and many children. Women are tied, usually, to having one child at a time—and in a more restricted period of their lives. But women also enjoy the advantage of knowing for sure who their offspring are, and can engage all their energies in the process of child-rearing with absolute confidence. Whereas men are far less likely to know without doubt who their closest kin may be, and therefore find it that much more difficult to bond with them.[198]

These differences alone are sufficient to have produced a distance between our sexes, who necessarily now have very different interests and genetic strategies. One sex has a large investment to protect and looks for quality and stability. The other has little to lose and tends to be far more interested in quantity and variety. So it pays males to be aggressive, hasty, fickle and undiscriminating. They pounce, they generally make the first moves and are more ardent in them. While it is more profitable for females to be coy, to find out as much as possible in advance and to wait and see what happens. They play hard to get and play for time by flirting. All moves with a sound grounding in evolutionary psychology.

Genes which allow females to be less inhibited leave fewer copies of themselves than genes which persuade them to remain highly selective. Among males, the best strategy is exactly the opposite one. The maximum advantage goes to those males with the fewest inhibitions. "Love 'em and leave 'em" is not so much a nasty piece of male chauvinist piggery as an accurate reflection of biological reality.

In a very real sense, each sex still finds it pays to use the other as a vital resource. Men are a little like selfish genes, looking for convenient vehicles to carry their inheritance into the next generation. Women are more cautious, like canny investors or developers, seeing men as inconvenient sources of a seminal substance that is nevertheless necessary to realize the potential of their precious nest eggs.

These bald descriptions sell both sexes short, but the two who differ so widely in interest and intent are bound to have different agendas and a conflict of interest. The fact that they manage to agree on anything at all is miraculous. Yet they do. Naked apes fall in love, form pair bonds and become sexually imprinted on a single partner—largely thanks to a radical change in the duration of female receptivity. Women still only ovulate at one point in their monthly cycles, but they can become sexually aroused at any time. The vast majority of matings in our species now have little or nothing to do with reproduction and everything to do with sociality and bonding. We are the sexiest of all the primates and though this may sometimes seem like a recent, decadent development, it is worth remembering that it has been this way for millions of years. It has had time to give rise not only to a splendid variety of courtship and precopulatory rituals and ceremonies, but to a mind-boggling array of sex-linked aberrations and pathologies.

The forbidden fruit was one of the sweet-and-sour variety, filled with the flavors of good and evil, sex as well as sin. And one of the consequences of having come up with something so pleasant and rewarding has been that some people will do almost anything to get more of it.

Agriculture helped. Farming introduced a new factor into a society of happy-go-lucky foragers. It produced what Lionel Tiger and Michael Fox (wonderfully apt names for a pair of life scientists concerned with our carnivorous heritage) have called "the great leap backward," separating us from the symmetries and satisfactions of the hunting-gathering way of life.[379]

Cultivation created a surplus of food, storable food, and this opened the way for one man, rather than the whole group, to gain control of such a store. And all of a sudden, over a relatively short span of time in historic terms, instead of being democratic sharers, some humans became autocratic owners. They accumulated sufficient resources to buy the favors of others, and if these resources were reliable enough, they came to need no favors in return—at least of other men.

Wealth, even of seeds or sheep, produced power. Power made it possible for some men to practice simple politics, to get others to do their bidding. And power, as Henry Kissinger once observed, is a great aphrodisiac. Men throughout human history have certainly been quick to treat power, not simply as an end in itself, but as a means to sexual and reproductive success.

Laura Betzig, one of a new breed of Darwinian historians, set out to discover whether human sexual adaptations have been exploited to give individuals a selective advantage—and discovered that this is one of our most predictable traits.[36] In all six of the great independent civilizations of early history, the rulers, always men, were despots who translated their power directly into extraordinary sexual productivity. That word "productivity," usually used in an industrial sense, is totally appropriate here. Each emperor established a carefully controlled breeding machine, designed and dedicated to nothing more than the rapid spread and dominance of his own genes.

Hammurabi of Babylon had thousands of slave "wives." Akhenaton, Egyptian pharaoh and husband of the gorgeous

Nefertiti, was driven nevertheless to recruit at least 317 concubines. Montezuma, the last Aztec ruler, enjoyed the favors of 4,000 young women. Several of the T'ang dynasty emperors in China demanded access to a minimum of 10,000 teenage girls. Udayama of India kept 16,000 consorts in palaces ringed by fire and guarded by eunuchs. And all of these rulers ran their gene machines in much the same way, recruiting prepubertal girls, pampering them under heavy guard, and servicing them as often as possible—sometimes even complaining of such onerous "duties."

The measures adopted certainly seem to bear out the claim of duty rather than pleasure, but in a survey of 104 other societies, Betzig found that even when such superpolygamy was not being practiced, there was always a direct correlation between power and sexual activity. She noted that: "In almost every case, power predicts the size of a man's harem." One hundred women for a small king, 1,000 for a great one. It is tempting to wonder where in this major league table to put American presidents.

Roman emperors made no secret of their prowess, though their biographers often rendered token disapproval. Julius Caesar's affairs were described as "extravagant," those of Tiberius as "worthy of an oriental tyrant." The wives of Claudius and Augustus collaborated in finding virgins to satisfy their husband's "criminal lusts." But Nero had to cater to his own peculiar desires, setting up a row of temporary brothels along the shore whenever he had occasion to leave home and travel on duty down the Tiber. And whenever there was an epidemic of young male slaves freed in Rome with surprisingly large endowments, it was widely accepted that these were the offspring of female slaves recognized as illegitimate sons by noble fathers forced to pay the price for following the example of their rulers.[37]

And if such practices are only to be expected of the "decadent" Orient or seen to be symptomatic of the decline and fall of ancient Rome, it is interesting to learn that sex and power were just as firmly entangled in Christendom. Census reports from France and England in the Middle Ages show a very heavy bias in favor of women working as employees in manor houses, châteaux, castles and even monasteries. Nominally, these were serving maids of various kinds, but contemporary accounts make it quite clear that they formed a loose sort of harem, in some cases even living in apartments affectionately known as *gynoecia,* a term now more commonly used to describe the female parts of a flower.[38]

The consequence of sexual despotism in all these instances is precisely the same. The genes of a minority, self-selected by wealth and position, spread rapidly through the population. Male power was being traded for female reproductive potential, giving each child the possible advantage of genetic friends in high places. There is even a name for it: *hypergyny,* the practice of mating upward on the social scale. It is usually limited to women, because they are more often treated by men as a limiting resource, as valued property. Men pursue and acquire, while women are protected and bartered. And powerful men can afford the most energetic pursuit and the sometimes high cost of acquisition. Those who have, get . . .

The result for poor, lower class, working men or people of peasant stock, however, was disastrous. In the Middle Ages they seldom married before middle age and sometimes never at all. And even when they did marry, it was often the right by custom of the lord of that manor to spend the wedding night with the bride. *Droit de seigneur* was a cunning, cost-free and biologically totally appropriate way of enhancing the lord's fitness. No community, as far as I know, ever practiced *droit de la madame.* Sleeping with the groom could only harm the lady of the manor's fitness. It was disastrous for Lady Chatterley and her lover. If a poor man ever had any premarital or extramarital sexual experience, it was more usually with a prostitute, or in other circumstances which carried the danger of disease and a further reduction in his health and situation. A "vicious" circle indeed—in the true sense of being both morally evil and injurious.

Female prostitution is a special kind of polygyny. An honest one in the sense that payment is made directly for an acknowledged service. When sex is sold, it is usually men who do the buying. The judgments which are made of the sellers, the ones vilified as "whores and harlots," are harsh, but not unexpectedly so from a biological point of view. One can almost hear the genes speaking in condemnation of those who abandon their valuable reproductive investment to strangers—unless the price is right.

On the Caribbean island of Dominica it often is. A large dark hummingbird—the purple-throated Carib—lives there in lowland dry forest, feeding on the nectar of trees which have an abundance of flowers. The males defend territories which include as many trees as possible, dashing between them like winged torpedoes, seeing rival males off with an aggressive ticking sound. They give smaller females the same abrupt treatment, unless such intruders produce the sort of signals that indicate a readiness to mate. If she does, he does, and after copulation the male lets the female feed undisturbed while he goes back to his perimeter patrol. Ornithologist Larry Wolf, who described this behavior, called his report: "Prostitution behavior in a tropical hummingbird," and was probably correct in doing so. What happens is that sexual favors are traded for reward, even if, unlike her human counterpart, the hummingbird female can still afford to trust the male and wait for payment later.[432]

It has to be said, though, that her gamble is not altogether altruistic or genetically unsound. If all goes well, she not only gets fed, but impregnated. And not just by any old bird, but by a male who is healthy enough and powerful enough to hold such a wealthy territory. She can afford to lose her virtue, because by winning control of the favored food sites, he has already proved that he is one of the best males around. This is not so much sin as good sense. An argument offered recently in controversy over the feature film *Indecent Proposal* by a large proportion of the audience who approved of Demi Moore's decision, in the role of a married woman, to spend one night with attractive bachelor Robert Redford in return for $1 million.

It is no accident that female prostitutes are more common than male ones. The difference is largely a matter of supply and demand. There are of course services for women who want to pay for sex, gigolos and escorts are reasonably easy to find, but most male prostitutes who ply for trade are interested mainly in male clients. And until the arrival of AIDS, male homosexuals were far more promiscuous than heterosexual males. Twenty-five percent of gay men interviewed in the San Francisco area in 1975 admitted to having had more than 1,000 partners.[71] The reason for behavior so excessive and dangerous may be that gay men have the same biological urges as straight men. Like all their sex, they put a premium on sexual variety, but also find themselves in gay company, in a world unfettered by female caution, with unprecedented opportunity to satisfy such desires. A little like foxes in a chicken pen. The analogy may be unflattering, but it is not inaccurate or uncommon. Overkill occurs when normal inhibitions are, for whatever reason, relaxed or suppressed. Then order falls prey to excess and behavior becomes pathological. In the post-AIDS period, something of very real concern.

Prostitution, while we are on the subject, raises another important issue. That of advertising in the form of sexual signals designed to attract a partner or a client.

In humans, mate selection is usually something a man does, or so we are led to believe. He can afford, in fact he is obliged, to take risks in the process. He has evolved to live dangerously, because being brave and "manly" increases his attractiveness and his chances of sexual conquest. It is no accident that we use military terms for sexual competition, or that military uniforms were once decorated with enough bright plumage to satisfy a bird of paradise. The object in each endeavor is to intimidate male rivals and to attract as many females as possible. Women are not supposed to live dangerously. Dressing provocatively is frowned on, sometimes even legislated against, not so much out of modesty as from a biological awareness that doing so could put a woman's life, and those of any children she may already have, at risk.

But a prostitute is, by definition, not interested in reproductive success. She can afford to live dangerously and dress in ways guaranteed to attract male sexual attention. That is her job and the good ones do it extraordinarily well, pushing the frontiers of what is acceptable, deliberately challenging all the dress codes, because that is the best way of advertising her trade. For her it pays to be outrageously conspicuous. It pays to advertise. But the rules and constraints are very different in the rest of the animal world, where it is the males who challenge taste and prudence, and are forced to such extremes by females who always seem happiest to invest in a big spender.

My first lesson in such elementary sexual economics came from a widow. Not she of the missing mite, but, as you will see, she of the many mites. I am talking about the African widow bird—so called because it bears a long, trailing, black tail reminiscent of a Victorian widow's funeral veil. It is the male bird who carries this train, four or five times the length of his body, flopping with measured wing-beats across the open grasslands each summer, flaunting his finery, making light of his considerable burden in the interests of attracting and holding the attention of a harem of dowdy brown females.

He is the peacock of perching birds, a natural strutter complete with crimson epaulets who can, if he works hard at it, hold a territory that includes perhaps half a dozen hens, each laying three speckled eggs in cunningly concealed nests of growing grass. It seemed obvious to me as a young and ardent bird-watcher in South Africa that the best-endowed males had the largest harems, while those whose feathers were depleted or damaged often failed to breed at all. I assumed that the harem masters were simply the biggest and most aggressive birds, those who had come out best in the battle with other males for breeding territory, but I was disabused of this idea when a freak accident gave one of these rivals in my home area supernormal sex appeal.

The road through the grassland near our farm was being paved, the weather was unusually warm for early spring and the tarmac lay in pools on the new surface, settling but not quite setting, looking for all the world like water in the heat mirage of early afternoon. I thought at first it was water and that a widow bird was bathing in it, until I realized he was stuck there foot and tail, struggling to be free from the sticky surface. I took great care not to damage his delicate toes, but a couple of central tail feathers were heavily tarred and torn from their follicles before I could release him completely. He flew away immediately, tarred and feathered as he was, apparently none the worse for wear otherwise, except that the loose feathers remained stuck to his tail, making him somewhat unwieldy, like an overburdened toy kite.

During the following weeks I saw him several times again, still struggling with the extended tail that was now perhaps six or seven times longer than his body, doing his best to compete against undamaged rivals, having a hard time of it, being blown astray with each gust of wind, sometimes even grounded. He had my sympathy, but this was hopelessly misplaced, because when nesting began he had accumulated the largest harem anyone in the area had ever seen. There must have been nearly twenty hens in his entourage by the time the codpiece and his act fell apart, but by then he had feathered a dozen nests with over thirty eggs and set up a gene pool big enough to bathe in.

The moral of the story is that nothing succeeds like excess. The male with the prosthetic tail may not have been the "best" male available, but he was certainly the most fashionable. This flamboyant individual was ill-equipped to compete with other males on all the usual terms, but so attractive and "well hung" in female eyes that there was no competition. The hens chose him *en masse,* showing a clear preference for his exaggerated tail, starting a despotic fashion that could have led, if further reinforced, to even more grotesque peacocklike lengths. Logically, such runaway selection ought to stop when males become so encumbered that they get eaten by grateful predators long before they get a chance to breed. But the evidence from this and other species suggests that another factor has been introduced. The very fact that these males are obviously handicapped by prodigal plumage, unrestrained colors or extraordinary growths, and still manage to survive, counts in their favor. And the more extravagant the handicap, the more costly it seems to be in terms of survival, the more attractive it becomes to females who see such expensive display as a signal of genetic quality.[437] The avian equivalent of a BMW.

The Swedish ornithologist Malte Andersson has recently provided more mathematical proof of such preference in a series of elegant experiments on widow birds in Kenya. He caught males in breeding plumage, cut some of their tails short and used Superglue to add the missing portions on the tails of others. He found that the number of mates each male attracted rose in direct proportion to the length of his tail. In this species at least, sexual success *can* be measured in inches.[9]

It probably once was in ours. Despite all their best endeavors, young women still tend to gain weight unevenly, accumulating fat mainly on their breasts and buttocks. And the reason they do seems to be the same as that of the widow bird. It is sexy. In simpler times, a man looking for a mate needed to be assured of two things—that her hips were suitable for bearing his children, and her breasts capable of feeding them. So men came to prefer women with relatively wide hips and large breasts, and women tried to please. If they didn't have such attributes, they made it look as though they did. Fat breasts of course don't produce more milk than lean ones, and fat hips are no further apart than lean ones with the same pelvic structure. But it worked. Men fell for it, and some still do.[247]

Nobody said advertising had to be honest, all it needs to be is conspicuous.

In 1982 William Hamilton—who has an uncanny knack of being at the center of biological debate—added to the stirs he had caused with kin selection in the sixties and reciprocity in the seventies, by introducing a new and startling idea. Sex, he reminded us, is unnecessary and irrelevant.[171] Lots of species manage very well without it. The average lawn never gets to set seed, and those simple souls who divide or clone pass on one hundred percent, not some miserable fraction, of their genetic material. The only real advantage sex provides is genetic variation. It mixes up the genes in a population, and one of the results of this unrest could be to keep potential parasites confused. Sex, suggests Hamilton, is in effect a kind of vaccination against disease. And if this is true, he predicted, the more a species is bothered by parasites, the more emphasis it will need to put on sexual selection. It will be forced to concentrate on finding and attracting a variety of mates with different genes that can keep it a step ahead of its invaders with each new generation. It will need to advertise, to pretend to be healthy. He predicted that the most flamboyant species would turn out to be the ones most prone to parasitism, and that within these species, the most conspicuous males would also prove to be the ones with the lowest parasite loads. Only they could afford the expense in time and energy of being extravagant.[172]

Hamilton and a colleague took a close look at "bright" birds, over 600 species of them, and found that the more parasites, the showier the species.[438] The same is true of tropical fish, both fresh-water and marine, and may even be true of humans in that men in polygynous situations with pathogen problems tend to dress for maximum effect. Think, for a moment, of the young pimp on a downtown street corner, in all his finery.[246]

In the time of AIDS, the idea of sex as a weapon *against* disease needs some getting used to. But historically and biologically, the evidence is persuasive. We are in a perpetual battle with parasites, and keeping them out means changing locks a little faster than they can change keys. Sex gives us a new advantage with every generation involved in the struggle, altering genetic combinations just fast enough to keep a step ahead. Often that is all that distinguishes a successful species from one that fails—an ability, like Lewis Carroll's Red Queen, to stay in the same place in the race by running as fast as possible.[321] Each victory is nothing more than a temporary respite. There is no end to the battle in sight, just as there is no perfect strategy that will bring evolution to its final destination. The struggle goes on.

Bear in mind, however, the fact that the competition is not between species. It is between individuals within the species. If two Maasai warriors are running from the same lion, they are not trying to outrun the lion. They know no man can do that. They are simply trying to outrun each other in a race to avoid being the one who gets eaten by the lion. Actually, being Maasai, a more likely outcome of such an incident is that the *moran* will stop and turn on the lion and start a different kind of competition. One designed to find out which of the two men is the greater warrior, best able to kill a lion and prove himself a worthy mate. Either way, in flight or fight, the winner improves his individual fitness and the chances of his genes being better represented in the tribal pool. Even if all this makes him in the end is a Maasai woman's insurance policy—not so much against lions as against the dangers of her children dying from infection by new strains of malaria or bilharzia.

In advertising, as in sex, there are two sides to every story. There are sellers and buyers, and buyers are admonished, with good reason, to beware. It is in the interest of the seller, particularly if the object of the sale is himself, to exaggerate the quality of the goods, to make light of a family history of congenital heart disease, or to hide altogether the fact that he may have parasites. A female buyer's best hope of making a good deal is to insist on a "test drive" or to arrange in some way to see the goods in action. The most effective strategy, the one which puts her at least risk, is to see him in action against another male, the seller of alternative goods. She needs to practice comparative shopping, and very often this is alarmingly easy to organize.

All biologists involved in field observation of behavior can offer examples of their own, but one of the best comes from an extensive study of the northern elephant seal made by a colorful man with a head start in male self-advertisement. He was christened Burney Le Boeuf. On many occasions he has watched subordinate male seals trying to mate surreptitiously with females while their larger rivals were sleeping. Such attempts rarely succeed. What usually happens is that the female protests loudly, squawking, wriggling, flipping sand in the air. "Her behavior activates the social hierarchy: it literally wakes up sleeping males and prompts them to live up to their social positions." What they do is come rushing to the source of the disturbance, not to the female's aid, but into a free-for-all battle with each other. The first result of this general alarm is that the young pretender is sent packing, but the most important consequence is that the male who wins the conflict—invariably the dominant bull—celebrates his victory by copulating with the scheming female. By inciting males to compete for her, she gets exactly what she wants—the best genes available.[83]

And if the best available don't seem quite good enough, she will often try to improve on them, even to the extent of overloading the system. Think of Eve beguiling Adam, of Salome seducing Herod into beheading John the Baptist, and of Macbeth consumed by his wife's ambition. It is a familiar scenario, with its roots in the demanding female role in mate selection, which continues to resonate with us today.

In the 1966 film of Elmore Leonard's *Hombre,* Diane Cilento relentlessly pushes the Paul Newman character to be more caring, more like herself, better than he can be, until in desperate circumstances she goes too far, goading him into hopeless action against a kidnapping bandit. Before going to his certain death, Newman dismisses Cilento's pretense that all she wants from him is a knife to free the hostage herself, and gives voice to everyman's exasperation in the face of everywoman's need: "No, lady. That's *not* all you want from me . . . "

Monogamy, apart from exceptional species such as the gibbons, is rare among mammals and practicing it sets us apart from nearly all our relatives. Literally. And sometimes this is just as well, because we all carry a little of the langur in us. Stepchildren are sixty-five times more likely to be murdered than children who still live with both of their natural parents.[94] A fear of new stepfathers is natural, reasonable and very hard to overcome. Chimpanzees solve the genuine problem of infanticide by confusing the issue—again, literally. Female chimps tend to be promiscuous, mating with as many males as possible, sharing paternity to the extent that no male knows which offspring are his, and treats them all with appropriate caution.

We, of course, are not chimpanzees. We are far too common, for one thing, and have become so because of some surprising adjustments to our social and sexual lives. We have reinvented monogamy, largely because women prefer it that way, showing a marked preference for monopolizing a man, even in polygamous societies. But instead of going into gibbonlike isolation, we live in large colonies where both sexes come into frequent contact with members of the opposite sex, provoking clandestine encounters, most often between high-ranking males and females of other ranks. Adultery has always happened and is the source of more prurient interest and hypocritical scandal than almost any other aspect of human behavior. But, Kinsey notwithstanding, it wasn't until the 1980s that anyone realized how common it was and how it colors our behavior and our inheritance.

Robin Baker is another British zoologist with a talent for turning the tables on tradition, asking new and interesting questions about things we tend to take for granted.[24] Women marry nice young guys, but have affairs with less scrupulous older ones; or they marry rich but ugly men and take handsome young lovers. Don't they? Even without the fictional stereotype of Madame Bovary, everyone knows someone like that who wants to have their cake and eat it too. It's only human . . .

Only it isn't. Most birds are monogamous and share the burdens of incubation and feeding their young. They make an appealing picture, dropping an endless supply of good things into nests full of gaping beaks, providing the images that go so well with talk of "nest eggs" and savings and investments in the future. But once genetic blood testing became reliable and economical enough to be used as a standard zoological tool, we learned to our considerable surprise that there were well-disguised cuckoos in almost every bird's nest, turning an extraordinary number of species into obligatory cuckolds. Most birds, it seems, are bastards, in every sense of the word.[39]

This discovery opened up a whole new line of research into what is now known as "sperm competition." It is a fascinating field, filled with sperm that race each other to the egg, with *kamikaze* sperm that form a plug to close the vagina door behind them, and with animals that dilute, scrape out or even wage chemical war on the sperm of rival males. None of which would be necessary if monogamy worked, if everyone was faithful to their mates. In birds they are quite clearly not. The best tactic for many females seems to be to mate with mediocre but reliable husbands, and have affairs with genetically more desirable male neighbors—usually someone else's husband. Monogamous barn swallows do it all the time, and go on doing so for the very good reason that it gives the males *more* young, and the females *better* young.[274]

Baker suspected that the same genetic logic applied to human adultery, and had the courage to look and to ask. The looking involved blood tests in London and Liverpool, both of which showed that twenty percent of children in apparently stable families were not in fact the sons or daughters of their ostensible fathers. That's one cuckoo in every five. Adultery was clearly involved and so Baker began to wonder whether sperm competition is something that happens in humans. He asked and found not only that it does, but that both men and women have wonderful mechanisms to deal with it.[22]

It seems that women have two types of orgasms. One, the most fertile, involves strong contractions that hold sperm far longer in the vagina after sex. The other occurs earlier, is less muscular and releases semen as easily as if no orgasm had taken place at all.

Faithful women have each type almost as often. Among unfaithful women, the frequency of high-retention orgasms goes up to over seventy percent with lovers and drops to less than forty percent with husbands—an advantage of almost two to one. And, by accident or design, the adulterous women tend to take lovers at those times of the month when they are most fertile, or at least feeling most sexy. Taken together, even if adulterous women make love at home twice as often as they do outside the marriage bed, they are still more likely to conceive a child by a lover than by their husbands. "The typical woman's pattern of infidelity and orgasm is exactly what you would expect to find if she were unconsciously trying to get pregnant from a lover while not leaving her husband."[321]

And human male biology is geared to defend against just such a tendency. Baker discovered that men whose wives have been with them all day ejaculate much smaller amounts than men who have been separated from their wives for the same length of time. Their bodies behave as if they are compensating unconsciously for the possibility that their wives may have been unfaithful in the interim, and are anxious to give their own sperm every advantage in a possibly competitive situation. If you cannot guard your mate all the time, at least have sex with her as often as possible—and don't be surprised if your mating is more ardent after a time of being apart. Absence makes the heart, and several other crucial body parts, grow fonder. While familiarity breeds, if not contempt, then certainly a degree of detachment . . . [23]

Studies on Israeli *kibbutzim*, where unrelated children grow up in very close proximity to each other, show that such boys and girls very seldom end up marrying one another.[104] It seems that, in our species at least, the inhibition against inbreeding depends not so much on an instinctive awareness of genetic relationship, but on a tendency not to be sexually attracted to those who one has come to accept as immediate family. In most circumstances this will be enough to keep apart those who are too closely related for genetic comfort. Oedipus was just unlucky. It could happen to anyone who doesn't have a culture that keeps careful track of such things. And it seems that Freud was wrong in suggesting that tribal taboos against incest were necessary to keep us from giving in to deeply repressed desires. The truth is that in most cultures prohibitions have not had to be strictly enforced, simply because hardly anyone seems to be really interested in having sex with mothers, brothers, fathers or daughters.

Ornithologist Stephen Emlen, in eleven years of intensive study on carefully banded bee-eaters in Kenya, has yet to come across a single example of mating between siblings or between parents and their immediate offspring.[332] The genetic imperative seems clear and clearly has considerable survival value. Studies on captive or domestic animals, where inbreeding is enforced, show that the chances of genetic defect or even death rise so rapidly that even fruit flies pushed into brother-sister matings die out in less than seven generations. Only Mayan and Egyptian royal families seem to have gone out of their way to break the taboo, but suspicion grows that even in their case, these brother-sister unions may well have been marriages of convenience to cover some judicious outbreeding behind the scenes. Cleopatra, at least, seems not to have suffered from any obvious physical or intellectual shortcomings.

One legacy of sexual evolution which does affect us all is a marked predisposition to certain kinds of aggression. No society has yet been discovered without a trace of sexual jealousy, and few pair bonds anywhere survive without a little of the creative tension it provides. Even ducks are not immune. Several marine species, such as the eider, form perennial pairs. These winter in company with other ducks, but in the summer on the breeding grounds they become more particular about the company they keep. It is the female who takes the initiative, becoming visibly annoyed when a strange drake approaches. She points her Grecian beak at the intruder, stretching her neck in his direction, then lifts her chin and makes a querulous, nagging sound. Her mate, whatever else he may be doing, feeding, dozing or preening, responds immediately to this incitement and attacks the stranger. It is such a deliberate, specific act, switched instantly to a new target the moment she redirects her displeasure, that it could have been choreographed by Jerome Robbins for young female Sharks picking targets for their men from the ranks of the invading Jets.[382]

Such behavior seems anomalous at first sight, in ducks or humans, showing women in an unsympathetic light. But ever since Darwin introduced the idea of sexual selection, suggesting that females played a major part in mate choice, and therefore in evolution, students of behavior have been alert to such an influence. Misunderstandings have arisen mainly from studies of courtship in captivity, where females have no choice at all. Giving one male uninterrupted access to a female who cannot get away, leaving the pair on their own without peer pressure or the restraints introduced by rivalry, can even be lethal. Laboratory rats and white-tailed deer in parks sometimes end up killing their mates, and the same unhappy consequences are not uncommon in human pairs isolated by living conditions which are just as unnatural for our species. Most murders are domestic, sometimes the result of overpopulation, but more often the product of a different kind of pressure—that of too few people. The Second Principle of Pathics cuts both ways.

Male on male aggression, from the scuffles of adolescent humans on street corners to the elephantine competition for mates on the beaches off the coast of California, is based on equally predictable biological disparity. Males that conquer other males leave more offspring than those conquered. And the difference between the victor and the vanquished can be immense. In Le Boeuf's study on the beaches of Año Nuevo, the top-ranking bull elephant seal inseminated twenty-five times as many females in the harem as a low-ranking male condemned to live on the fringes of the colony. Given the size of the payoff, the level of such competition is inevitably high and begins surprisingly early. Even newborn male pups get in on the act, stealing milk from mothers other than their own, risking injury or death as a result of such theft, but also growing faster than female pups, who show no such larcenous inclinations. The well-fed young "double-mother-suckers" are more likely to survive their first winter and to grow big and strong enough to become beach-masters themselves.[238]

The parallels with human behavior are obvious. Boys in most cultures are more venturesome and confrontational than girls. From the beginning of social play at the age of two, boys are more aggressive in both words and actions. They tend to have more, and more explicit, hostile fantasies. Their play far more often includes elaborate threat displays, mock fighting and overt physical attacks that are directed mainly at other boys. Girls, on the other hand, are predisposed to be more intimately sociable and less adventurous. Even as newborn babies, they smile more than boys and go on doing so throughout their lives.[429] Part of our upbringing involves conditioning designed to exaggerate such tendencies, but the differences seem to be real and largely genetic. They follow the same distinctive pattern even in cultures, such as the San Bushmen, where children are reared without regard for their sex.[248]

Such biological bias is basic and necessary. Men are destined, as low investors, to compete with other men for access to women, and to do so more often and more vigorously than women need to compete for access to men. Women enjoy a high probability of reproductive success and do not need to indulge in risky behavior. While men, faced with massive differences between the rewards of success and failure, are condemned to be far more competitive and aggressive. The difference lies largely in our genes, or at least in genetically controlled sex hormones—something given traumatic demonstration by a medical disaster in the 1950s.

For a while, in the United States at least, it was accepted practice to treat pregnant women at risk of miscarriage with *progestin*, an artificial hormone. This was designed to mimic progesterone, the female sex hormone, but in some bodies it lost direction and turned out testosterone instead. When this happened, and the fetus was female, masculinization took place and a number of girl babies were born with pseudo-male genitalia, in a state of hyena-like confusion. Their labia were closed in a sort of scrotal sac and the clitoris enlarged enough to resemble a penis. These external manifestations were surgically corrected and the girls "trained" to think and act like ordinary young women, largely to no avail. Most of them showed a greater than normal interest in athletic skills, a preference for guns over dolls, slacks over dresses. And all came to be regarded as "tomboys," a fascinating label used since Shakespearean times to describe girls thought to be immodest, bold and boisterous—"hoydens who trespass against the delicacy of their sex."[319]

Studies on these women, and on those who were exposed prenatally to the same hormone but showed no visible external changes, found that all were more dominant, more aggressive, more adventurous than their classmates. Lacking access to the female hyena solution of pack leadership and full expression of their hormonal makeup, few were happy being women, and some distinctly dissatisfied with being assigned a female role in human life. It seems clear from this accident that sexual differences and divisions of labor are not just cultural artefacts. They are biological in origin and subject to very simple biological disturbance. Evolution has ensured that we are largely what we need to be—and what most men need to be, it seems, is appropriately aggressive. Enough at least to take advantage of the usual opportunities on offer. It is when the normal balance is disturbed biochemically, spatially or mathematically that pathological distortions occur.

When Fletcher Christian and his eight fellow mutineers from HMS *Bounty* reached Pitcairn Island in 1790, they were accompanied by six male and thirteen female Polynesians. Together, these fifteen men and thirteen women tried to found a stable colony and build a new life, putting themselves completely out of touch with the rest of the world by literally burning their boats. When they were rediscovered by American whalers in 1808, only one of the men was still alive, surrounded by ten women. The rest had been murdered. "The survivor was simply the last man left standing in an orgy of violence motivated entirely by sexual competition."[321] He was one of the Polynesians, who confessed, repented, made a strategic conversion to Christianity, and prescribed monogamy as the best solution for Pitcairn society. The present population of around a hundred people still swear by it and live by selling postage stamps to passing ships, but it is tempting to wonder what might have happened if one of the men in the original party had been replaced by a woman. The difference between stability and instability, between good and evil, can often be that small.

Being good and being bad are simply part of being human. We are products of our past and none of our direct ancestors died childless. Every one of them left us something to remember them by, and the sum total of this inheritance is what drives us. It makes us love and hate, it produces ambition and greed, homicide, suicide and envy; but it also results in charity, language, music and art. Our cultures are breeding grounds for inequity, but all also contain the seeds of their own remedies. We have, I suggest, an innate predisposition towards equilibrium. We recognize asymmetry for what it is, the natural state of things, and cultivate its qualities where they work to our advantage. We have become very good at reading the symptoms of impending imbalance in one another, and of acting to inhibit their extremes. Our societies have encoded such instincts in rules and laws, describing and defining what works and what does not, what is moral and immoral, good and evil. But the most effective restraints remain those which are built into us and come floating to the surface as required, quenching the flames of social fire.

Nature may be "red in tooth and claw," but it is also realistic. The struggle for survival is real and undeniable and we can be brutal when involved in a conflict of interest, or when the system of restraint breaks down—as it has most recently in human Rwanda. But we can also be astonishingly gentle and cooperative, skilled in the delicate art of reconciliation, buffering potential discord before it even becomes evident. We may be the toolmakers, the ones who turned the biological arms race into all-out war, but we are also the peacemakers, astonishingly and inherently skilled in reconciliation. So skilled that it is often easy to misunderstand what is going on, to condemn what looks bad without waiting to see how it works towards a greater good. We have become very fond of the idea of a "good" war, where it happens to serve our purpose, but it is only just beginning to dawn on those whose task it is to identify and encourage moral behavior, that there is such a thing as a "bad" peace—like the one the American Marines tried to bring about in Somalia, and the one the British army has been trying to make for so long in Northern Ireland.

Human "peacekeepers" have a great deal to learn from our relatives the chimpanzees, even those still living in Rwanda.

Chimps resemble us, or we them, in far more ways than we differ. They are thinking beings with prodigious memories, well capable of planning ahead. Most of this rational intelligence is deployed in, and may have evolved from, social situations. They are extremely skilled in the niceties of politics and image-making, of one-upmanship and reconciliation. Chimps know a good deal when they see it. As a zoo director, I soon learned that the best way of taking something harmful from a chimp was to trade for it. Disturbed and thoughtless zoo visitors sometimes threw broken glass or even knives and scissors into animal areas and, whenever this involved the chimpanzees, all I had to do was to take an ice-cream down and point to it and the object I wanted in return. Chimps are very quick to get the point, and our awareness of such intricacies in their behavior has soared during the last two decades, partly from extensive field studies, but most particularly from intensive observation, by day or night, of a colony at Arnhem Zoo in Holland.[398]

"The law of the jungle does not apply to chimpanzees," says Dutch primatologist Frans de Waal. But the politics of the playground or the street corner seem to fit very well. Consider the following passage and try to decide whether it applies to human or ape behavior.

> Everybody pulls strings. When a fight breaks out, others hurry over to watch, giving cries of encouragement. Some try to interfere on behalf of their favorites. Coalitions spring up and range in size from two to ten aggressors, but victims also receive help, leading to larger-scale confrontations. Combatants actively recruit support, drawing attention to their situation by screaming at the top of their lungs; they put an arm around a friend's shoulder to get him or her to join in; they beg for help from bystanders; they flee to an older or stronger protector and, from such safety, shout and gesticulate at their opponents.

This is in fact a passage from de Waal's notes on chimpanzee conflict, edited only to remove clues to the identity of the species concerned. But it could equally easily have come from a work by Jean Piaget on the origin of aggression in the human child.

Observations in Arnhem show that the end of a fight, or any

kind of aggressive encounter, is not the end of the affair. Revenge is an occasional result, but far more often the participants are involved in elaborate reconciliation. The opponents contact each other soon afterward, making the distinctive gesture with outstretched arm and open hand which chimpanzees use to beg for contact. They show more than usual eye contact, soft yelping, hand nuzzling and mouth kissing, and persist with these overtures until they are reciprocated and reconciliation occurs. Failing which, they may solicit contact with a third party, which involves more embraces than kisses and seems directly comparable to human consolation. De Waal, with justification, draws a nice parallel between one such incident involving recent chimpanzee rivals, and television news coverage of the meeting of the prelates of the Argentinian and British branches of the Catholic Church, who arranged a joint Papal Mass after the brief war over the Falkland Islands and exchanged an elaborate gesture of official conciliation in the form of a "kiss of peace."

The only time such overtures seem to fail for chimpanzees is when combatants are involved in some long-term dispute such as a power struggle for dominance within the group. Then the future dominant animal turns his back and walks away, and continues with such rejection until the eventual subordinate shows submission and deference by panting grunts and elaborate deep bowing movements. These are quite clear in their intent and instantly recognizable for what they are to anyone who has watched what happens at a Chinese political rally or in a Japanese boardroom. They are patterns of social integration which carry a powerful message of conditioned reassurance: "Everything is going to be fine, as long as you continue to show that you know your place." In any military hierarchy, all you have to do is give the expected salute to your superiors. We and chimpanzees seem to share the same fear of disturbance to our relationships and are equally sensitive to all the small signs that warn of an impending breakdown. They often cause us more concern than outright physical aggression, which is usually much easier to read and far more simple to deal with.

To buffer against unexpected breakdowns in their carefully contrived society, chimpanzees, like humans, establish coalitions reinforced by social bonds such as sitting together and mutual grooming. These bipartite relationships become integrated into a larger support network that has to be maintained by more elaborate, more overtly political maneuvers. And here, in chimps as well as ourselves, male bonds are more flexible, more prone to strategic rearrangement than the bonds between old-established pairs of female friends. Toshisada Nishida, working on wild chimps in the Mahale Mountains of Tanzania, reports on the tactics of an old male who regularly changed sides between two younger males, each of whom needed his support to dominate the other. Nishida describes this as "allegiance fickleness" and as power broking paid for by sexual privileges, and finds wry correspondence between such behavior and that of politicians involved in setting up fragile coalition governments with the help of still-influential older statesmen—a practice recently much in evidence in Japan.[287]

The ability to do this relies on the chimpanzees being able to make a subtle distinction between formal rank and real power. Rank is simply the position of an individual in the group hierarchy, one maintained and expressed by ritual gestures such as hair-raising or bowing, depending on your status, every time two chimps meet. These numerous "agonistic" encounters were once interpreted as evidence for high levels of aggression within the society, but are now seen to be deliberate and conscientious gestures designed to inhibit and control aggression. A kind of formality, a pattern of manners more than anything else, which is an essential requirement for relaxed relationships. It is part of the cement in a society in which true influence is more often to be found, though far less easily seen, in the manipulations of older, wiser power brokers and king-makers behind the scenes. Machiavelli would have loved it.

Female chimps are less likely to be fickle. Their coalitions are long lasting and relatively independent of status and kinship. They seem to be more concerned with establishing and keeping a set of solid relationships with a small, selective circle of friends, and a few more clearly defined enemies. They make friends and forgive enemies less easily than the males. "Over the years," says de Waal, "I have gained the impression that each female in the Arnhem colony has one or two absolute enemies, with whom reconciliation is simply out of the question." He points out that, in many instances of apparent male aggression, which were once scored that way, closer examination of the context reveals that the action was actually instigated by a female in pursuit of a long-standing grudge of her own.[399]

Pious hopes aside, it seems unlikely that in any species there can be one quarrelsome sex and one peaceful sex. The more we study other species, and the better able we are to understand our own, the more likely it becomes that competition is not the prerogative of either sex. Both compete intensely, but perhaps for different reasons. Males compete largely for mates or for the prestige that gives them access to more mates, and females compete for food and in support of their offspring. If there is a major difference, it lies in the nature of the "buddy system," a tendency to bond into all-male groups whose clearly defined hierarchy is established by aggression, but which works in the end to make rivalry more predictable and less divisive. Male chimps and humans are more inclined to physical violence. We are the obvious troublemakers, competing most overtly, but also finding a sort of unity in such competition. While, as Lillian Rubin points out, female struggles for their goals are rarely as clearly acknowledged or as simply resolved, leading more often to long-standing divisions as individuals do their best to stay away from rivals and from sources of competitiveness.[328]

Frans de Waal, over fifteen years of intensive observation of the chimpanzee colony at Arnhem, has provided what is certainly the most detailed and revealing study ever made of a political struggle in another species. He has recorded and documented every stage of a shifting dance in which three adult males try out all the permutations available to them in an attempt to arrive at a stable resolution. The fact that they ultimately fail and that one of them dies in the attempt does nothing to detract from the subtlety and skill of the performance. And that it turns out to be a tragedy has little, it seems, to do with the captive situation. Recent evidence suggests that what happened is not unusual in the wild. It is an epic tale of passion and intrigue, of argument and conciliation, with psychological and social insights worthy of Dostoyevsky.

This is *The Brothers Karamazov* written for chimpanzees.

The story involves a scheming elder, himself an ex-group leader, and two younger rivals to the throne. All three associate intensively, usually in pairs, with the excluded individual doing everything he can to prevent the other two from getting too close. Divide and rule is the rule, supplemented by a basic awareness that "strength is weakness," that to be too strong is to invite others to cooperate against you. For years the triangle turns, always coming to rest with its apex up, with one of the trio dominant, but only allowed to be so by the complicity of the other two. Leadership, sometimes by design, sometimes by default. This is the story of all triadic coalitions.

In the last months, a stability of sorts settled on the colony, with an undercurrent of unease produced by the fact that the Machiavellian older chimp seemed to have decided that his best hope, his greatest chance of enjoying continued access to the favors of the colony's females in the long run, rested in an alliance with the less able and more pliant of the two younger chimps. There were moments, de Waal admits in retrospect, that should have given cause for alarm—when the more able young chimp was putatively in charge, and being submitted to with elaborate ceremony by the other two, but clearly mistrusted them. However, heads that bear the crown are expected to be a little uneasy, and the human watchers, though they noticed that his usual regal posture was being replaced by an uncharacteristic fetal huddle, assumed that nothing more serious was in the air than another change of government, and took no evasive action.

During the night of September 12, 1980, the two conspirators crept up, like Cassius and Brutus, on their sleeping leader and disabled him by biting off his testicles. That wasn't the end of the carnage, which left the victim with wounds on his head, flanks and back, four missing toes, and several fingernails torn loose. He was found the following morning, still alive in a pool of blood, but soon died of shock and trauma, leaving a silent and chastened colony desperately grooming each other to iron out the tension. Their usual noisy activity only resumed after the corpse had been carried away.

De Waal, a veteran ethologist with wide experience of animal behavior, admits that objectivity in the face of the killing was impossible. He believed that the old campaigner was to blame. "He was, and still is, the one who decides everything in the chimpanzee colony. His partner, ten years younger, seemed only a pawn in his games. I found myself fighting this moral judgment, but to this day I cannot look at him without seeing a murderer."[399]

De Waal left Arnhem soon afterward for a primate center in Wisconsin, where he continues to monitor a stream of new discoveries from the wild which reveal an unexpectedly dark side to ape nature. Chimpanzees are now known to hunt and kill baboons, bush pigs and small antelope, eating these whenever they can. Recent film of a group of chimps cooperating in the ambush and dismemberment of a leaf-eating colobus monkey is frightening because it reveals for the first time the transformation of our cute cousins into a howling, bloodthirsty lynch mob. During the last twenty years, almost as though there had been a sea change in the psychology of the larger primates, field workers have been finding new evidence of violence in their lives. Chimps occasionally kill and eat the infants belonging to other bands.[57] After the breakup of her study community at Gombe in Tanzania, Jane Goodall reported that males from the larger splinter group were sometimes seen to mount extended and brutal attacks on the other group, involving "a swift, silent approach through the undergrowth" of several males who descend on an isolated victim and pound, hit, trample and bite him to death, on one occasion apparently even castrating him in the process.[152]

This particular mutilation is chilling in the light of its resemblance to the practice, during some human wars, of emasculating the enemy. Those accused of plotting to overthrow the government of Surinam in 1982 were executed and ritually castrated. There were similar reports in 1993 of prisoners in Bosnia and Somalia being deliberately disfigured in the same way, some of them before dying. And the Mafia are said to send potent messages to those likely to molest their women by making an example of one such suspect every now and then, leaving him with all his severed genitals stuffed into his mouth.

It is possible, sadly, to believe such acts of human degradation.

We are becoming inured to almost anything our species might do. But the possibility that such behavior could have biological roots makes me queasy. I am also disturbed and intrigued by one very strange aspect of an attack reported by Goodall. It involved several adult male chimps who came across a female from another group with her infant on the edge of their territory. She did what any female chimpanzee will do in such circumstances, made submissive sounds and reached out gently to touch one of the males in reassurance. His response was extraordinary. He not only rebuffed her overture and moved quickly away out of reach, but performed a sort of exorcism, an act of real and ritual purification very reminiscent of Pontius Pilate. He picked a handful of leaves and vigorously scrubbed his fur precisely where she had touched him.[153]

This behavior is a crucial clue to what happened next. All the males, acting in concert, surrounded the female, attacked her brutally, seized her infant and killed it. This incident shocked the human field workers at Gombe, because the attackers and their victim had once been close as members of a single community. "By separating themselves," said Jane Goodall, "they forfeited the right to be treated as group members—instead they were treated as strangers." De Waal, fresh from the traumas in Arnhem, went further, suggesting that the behavior was extreme, not simply because the participants were now strangers, but because they *had once been friends.*[399]

That adds another dimension to the encounter, one that will resonate with anyone who has been through an acrimonious divorce or remains part of a family split by feud. Aggression against an outsider is part of business as usual, something genetic, not just condoned but expected, a sort of "weak evil." But the introduction of previous relationship, which is not something humans or chimpanzees easily forget, adds a sense of betrayal to the mix, making it possible, often inevitable, that things will become very ugly indeed—part of the awesome pattern of "strong evil." The tendency for civil wars to be more vicious, more sadistic, than wars fought between strangers is well known. And Yugoslavia and Rwanda ought to be potent reminders to those who had forgotten.

Perhaps the chimp who tried to "wash his hands" of the affair at

Gombe was aware, at some level, of this danger and made an attempt to avoid such escalation. In the event, he failed, but the attempt is at least encouraging, suggesting that not just evil is rooted in our tissues, but perhaps also a nascent sense of the idea of good.

Through most of human history we lived in small groups, less than 200 strong. Most of those we lived among were relatives, our flesh more or less. What happened to them, happened to us. What we did unto others, we did to ourselves. Altruism made sense, and continued to do so until we moved more and to greater lengths. Until we found ourselves amongst strangers, people who were no longer family—and therefore not *familiar.* Then the rules changed and became something quite different. Rules inside the family are easy, the genes take care of that. But rules outside the family are hard to find and keep. They deal not with what *is,* but with what *ought* to be. They set up ideals and create the need for ethics, which viewed at its most fundamental level is perhaps nothing more than a symbol of fear. Fear of the unknown, but also of what we do know. Fear, like that of the chimp on the edge of his territory and on the brink of losing control, that we can no longer trust ourselves to do what is right and necessary.

One of the obvious functions of bonding which takes place between males in groups is that of territorial defense. Intergroup violence seems to be one of the legacies of being an ape. In the case of the male chimps involved in the Gombe incident, the intruder was not a rival male, who would have been dispatched with the usual military precision, but a female they once knew, whose offspring now carried few of their genes. The infanticide was a genetic imperative, but the identity of the female was a source of almost unbearable confusion. The secret of stability in most situations is the existence of a clear organizational structure, of a hierarchy or kinship or coalition which provides the necessary framework for decision. Something that gives the rules easy meaning and obvious application. Nobody likes social or political confusion.

De Waal once spent several hundred hours monitoring an ongoing dominance struggle between two of the three rivals in his chimpanzee triad. "My objective," he says, "was to be there when one of the two males would formally submit to the other by means of the familiar bowing and pant-grunting ritual, and to compare the relationship before and after this moment. After three months of daily intimidation and noisy conflict, one finally capitulated. I do believe I witnessed his very first submissive grunts, because the other chimpanzees responded by rushing to the two males to embrace them both. The group must have been waiting as eagerly as I had."[439]

Submission, appeasement and *rapprochement* are part of aggressive behavior, and frequently lead to integration. They provide the necessary preconditions for peace and conflict resolution. But peace in itself is seldom a goal for such behavior. The maintenance of relationship nearly always is. We will do whatever it takes to keep an important network intact: lie, deceive, steal and even kill, everything except "lose face." The expression itself is interesting for a species whose primary system of communication concentrates on the facial area. Loss of face is the ultimate calamity for members of any such social species in which survival depends upon acceptance, and acceptance on rank, and rank on success of being who you seem to be. So it is not surprising that individuals sometimes go to extraordinary lengths to find solutions that obscure the truth while making it possible to "save face."

Primates are very skilled in obfuscation of this kind. Monkeys and baboons feign elaborate interest in nonexistent objects or events, looking everywhere but at the source of embarrassment, hoping it will go away. Chimpanzees sometimes embroider these charades by hooting loudly, summoning up the assistance of third parties, creating diversions which allow two adversaries to settle their dispute without either having to back down. The anthropologist Colin Turnbull describes in detail one of the most delicate and inspired maneuvers of this kind which he observed while living amongst the Ba'Mbuti pygmy people of the Ituri forests.

Housing there is simple, consisting of huts of sticks and leaves built by and "owned" by the women, who sometimes use their property as a ploy in matrimonial disputes. They will, when pressed, begin methodically to dismantle the family home, until the husband relents and stops them before they go too far. On this occasion, however, the man was unusually stubborn and waited until all the leaves had been removed before telling the whole camp that his wife was going to suffer from the cold that night. That marked the point of no return. She was compelled, with tears running down her cheeks, to continue the demolition; while he sat silently, hugging his knees, looking just as miserable, but unable to stop her without one of them losing face. "By this time," says Turnbull, "the camp was agog, because it had been a long time since anyone had seen a domestic dispute carried quite so far."

That was when inspiration struck. "He brightened up suddenly, turning around to see how far the demolition had gone. Only a few sticks had been pulled out, and he called to his wife and told her not to bother with the sticks, that it was only the leaves that were dirty." She looked briefly puzzled but then, understanding, asked him to help and the two very gravely washed every single leaf in a nearby stream and then put them joyfully back on the hut. Nobody in the camp believed the pretense, but everyone was glad the quarrel was over, and for several days they politely talked about the problem of dirty leaves, even taking a token one or two down to the stream to wash in solidarity—as if this was a perfectly normal procedure. But, says Turnbull, "I have never seen it done before or since."[389]

Compromise works, leaving no winners or losers. The art of the fair draw. It is a sophisticated strategy that differs from reconciliation in that the cause of the dispute is never even examined, at least not overtly. But it is just as much part of aggressive behavior as fighting or running away. Konrad Lorenz, the Nobel Prize–winning German ethologist, was right when he pointed out that: "Aggression can certainly exist without its counterpart, love, but conversely there is no love without aggression."[245] The kiss is just an inhibited bite. And the Seville Statement on Violence, which was drafted by an international committee of scholars in 1986, was probably misguided in affirming that: "*It is scientifically incorrect* to say that we have inherited a tendency to make war from our animal ancestors; or that war or any other violent behavior is genetically programmed into our human nature; or that in the course of human evolution there has been a selection for aggressive behavior; or that humans have a 'violent brain'; or that war is caused by 'instinct' or any single motivation."[392]

The statement from Seville closes with the optimistic suggestion that: "The same species who invented war is capable of inventing peace," but the truth about evolution is that we cannot claim the blame or the credit for either. War-waging and peacemaking are as old as ants and apes. There is no point in downplaying the reality of genetic inheritance or trying to mask the extent of continuity between ourselves and other animals. The picture which emerges from recent research is not one of continuous strife or cutthroat competition in nature. Most species lead congenial lives and have long-term investments which they protect by exercising impressive constraint. There is widespread recognition that the greatest threat to survival comes not from outside, but from within our societies. Forget about great white sharks and killer bees. We have most to fear from our own species. It was the arrival of Man Friday on his island that really tested the survival skills of Robinson Crusoe. And conversely, it was the practice of social awareness which provided the selective pressure necessary to give us our much-prized big brains.[401]

There is abundant evidence to show that something like moral behavior exists in other species. Chimpanzees presented with a situation in which there is too little of a prized food to go round tend to share what there is. When faced with such a dilemma, they show a hundredfold increase in kissing, touching and embracing—the kind of contacts likely to appease potential aggression. And aggression, when it does appear in this context, is far more likely to be "moralistic" in the sense that it takes the form of sanctions against individuals who are reluctant to share.[400] Sharing under pressure may not seem exactly gracious, but it provides a good example of the adaptive, prosocial uses to which aggression can be put. It is as wrong to see aggression as strictly antagonistic as it is to view sex as purely reproductive. Each behavior flows over and into the other in ways that become more complex as evolution goes on, providing a constant source of surprise to those still resistant to biological attempts to explain human behavior.

The facts are simple. Societies, both human and animal, are networks of interacting individuals. What one individual does affects the others, and social behavior is very largely geared to improving the cost-benefit ratio of such interactions. Individuals and societies compete for the solution most cost-effective to them. Strategies vary with circumstances. Sometimes cooperation turns out to be the most effective way to compete. Sometimes not. And it doesn't seem to matter too much whether the society in question is a chimp colony or a nation-state. It is not ideology, but self-interest that keeps alliances together.[125]

After the defeat of Napoleon, the new political order in early nineteenth-century Europe was established by the Holy Alliance, a marriage of convenience between Tsar Alexander I of Russia, Emperor Franz I of Austria, and King Friedrich Wilhelm of Prussia. The union of these three monarchs, who promised solemnly to "treat each other as brothers, support each other whenever possible, and treat their subjects as fathers treat their children" was dismissed out of hand by the British as "mystic nonsense." But it worked surprisingly well because, despite their differences, all three men were dedicated to the restoration of the old order, which meant the survival of their individual thrones. By 1907 the situation was very different. Russia had been defeated in the Far East by Japan and was happy to join with Britain and France in the Triple Entente, unified only by their common fear of a newly rich and restless Germany. And thirty years later the dance began yet again, with Germany rearming, France allying itself to Czechoslovakia, Hitler and Mussolini becoming cozy, Stalin signing an astonishing German-Soviet nonaggression pact, and Churchill vowing to enter into a deal with the devil himself if it would stop the Nazi advance. Functionally, there is little to distinguish this self-serving merry-go-round of opportunism from the politicking that took place between those chimpanzee rivals at the Arnhem Zoo.[174]

When it comes to the management of actual conflict, there are many more similarities. Both humans and animals show xenophobic behavior, beginning as simple genetic tendencies to be unfriendly to strangers, ranging through feuding against and raiding on neighbors, to major lethal encounters. As the size of the conflict grows, both show the same disposition to form larger and larger coalitions, drawing on progressively wider relationships. And both temper their aggression with well-defined and clearly understood restraints. The roots of war lie deep in nature, it seems, but then so too do the roots of peace.[44]

Sex may be central to our lives, but it is restrained, both by our biology and by our society. Sex is also asymmetric, which leads to sexual differences in strategy and priority. And these too can be divisive, but are in their turn constrained and redirected, making sex as important for bonding and sociality as it obviously always has been for reproduction.

Sexual competition lies at the heart of natural selection and encourages excess, rewarding those who most successfully promote their wares. Advertising pays, but it is all too easy to let it get out of hand. Aggression too is a part of such competition, growing out of a basic biological bias that leads, not just to adultery, but to jealousy. And this too can spin out of control. We have learned to live with these divisions, finding ways of taming and banking the fires, refining aggression in ways that are far from antagonistic. We tolerate and encourage *appropriate* aggression, sharing with our closest relatives an array of peacemaking and peacekeeping devices which is truly astonishing. We have, by taking thought and seeking balance, managed to make a virtue out of a vice, turning aggression into something productive and necessary—a "weak evil," when we do it right.

Being bad can be good. Even language now allows that distinction. But there are times when: "Things fall apart; the center cannot hold; mere anarchy is loosed upon the world."[435] Then something very different appears, something far more difficult to understand. It isn't pretty and it isn't easy to define in biological terms, but there seems to be a lot of it about right now, busy slouching its way into our lives.

The time has come to turn that way. We have everything we need now to take a close look at "strong evil."

FOUR

THE EVIL THAT MEN DO: THE ANTHROPOLOGY OF EVIL

Eating people is wrong. Right? Yes, but some of my best friends are cannibals.

They live on the Casuarina Coast, the delta area of Irian in Indonesian New Guinea. It is a flat mangrove swampland, a place of water, mud and lush vegetation. An area ruled by rain and tide. At low tide, mud banks extend for miles out into the sea and at high tide the sea runs right into the forest. It is impossible to say where water ends and land begins. The people who live there don't even try; they have come to terms with their environment, floating to and fro on the tide in their dugout canoes, or perched above it in their houses on stilts. But they have a very clearly defined notion of their own identity.

There are about 20,000 of them and they call themselves the Asmat, which means "the people—the human beings." All others everywhere, all outsiders, are known very simply as Manowe—"the edible ones." And if I feel no threat in this, and count a small number of individual Asmat as true friends, it is simply because I have not yet been placed in a situation amongst them in which it has become necessary for them to transfer me to the active list of those in direct danger of having their heads hunted.

This tendency to classify, to divide the world into "us" and "them," into members versus nonmembers, friend or foe, is one of the few true human universals. Something common to all people everywhere. *Homo sapiens* is considered as a single species in biological and taxonomic terms. We are, but this tendency to consider one's own group, the in-group, as different and superior, is so widespread that it creates a need to see ourselves, in cultural terms at least, as different species. The German ethologist Eibl Eibesfeldt calls this *pseudospeciation* and suggests that denying our enemies or opponents a share in our common humanity puts all conflict between clearly defined groups into a new behavioral category. We ought to see such struggles as taking place at the interspecific level. Those involved certainly do.[117]

The "gentle" Bushmen of the Kalahari use one of their explosive clicking sounds, taking at the same time great delight in the difficulty outsiders have in doing so, when they call themselves! Kung—the humans. Others to them are essentially nonhuman, different and, by implication, inferior in several important ways. The normal rules of humanity just don't apply to them. The Mundurucú headhunters of the Amazon add to such exclusivity the deliberate and powerful ploy of turning outsiders into animals. Their warriors speak of *pariwat* or "fair game," putting enemies into the same functional category as peccary or tapir, beings it is permitted to hunt simply because they are dangerous *animals.*[281]

In both Irian and the Amazon, headhunting takes place against the background of a shortage of high-quality protein. Resources are limited, despite the tropical setting, and competition for them is strong. Both people have elaborate systems of territoriality and food-sharing, most based on rituals designed to foster conservation and good ecology. Waste is one of their cardinal sins, as is a lack of generosity. Their cultures are products very largely of their environment. They encourage acts which are good for the ecology, and discourage acts which are bad. And those which are truly ecologically unsound are translated very easily into what they consider as "evil."

Headhunting isn't one of these. It is a good and necessary act to begin with, because it reduces competition on the hunting and fishing grounds. But it is also dangerous, both physically and emotionally. It carries the risk of retribution and of responsibility for the spirit of the deceased. Even for a cannibal, killing isn't easy. So it has to be hedged about with ritual and restraint, and given social and religious dispensation. It has to be justified, and one of the easiest ways to do so is to find some way of making the enemy unworthy, giving him an inferior or unclean status. Turning him from one of us, the human beings, to one of them, those less than human. Those who clearly *deserve* to die.

That is the Mundurucú solution, but the Asmat answer is far more creative. They have turned population dynamics into an intricate and strangely beautiful game, setting up one of the most elaborate ecological adaptations I have ever seen. They eat each other, happily.

It sounds simple, but headhunting as it is practiced in Irian is not simple and has nothing to do with war. It is the formal and ritual expression of a perceived need to keep things in balance. It is an Asmat admission of human responsibility for human destiny, a radical but nevertheless highly realistic solution to the problem of overpopulation. And whatever else you may feel about it taking place, you have to admit that it works.

It begins with a tacit acceptance of death and the necessity for a succession of generations. "The man is dead," they say, "long live the child." When an Asmat comes of age and is ready for initiation, he needs two things—a giant canoe with an extravagantly carved prow, and a special name which falls into the category of *owom*, something secret and special. The canoe is easy, every village owns one and it merely has to be spruced up and decorated with vegetable dyes and lime. But the name has to be imported, and that is not so easy.

Every Asmat child is taken soon after birth to a "seer," who examines it carefully and names the spirit which inhabits the tiny body. That "spirit name" is used, almost exclusively, until puberty when children are considered to have acquired some degree of individuality and can be given a "human name," which derives usually from the father or grandfather and is taken very seriously. It is a name with power and responsibility, one known to the community, but never used lightly in front of strangers. To protect the sanctity of the human name, everyone also has a "nickname" that refers to a title of honor or, more often, to an incident or characteristic that gives other people pleasure.

Among my Asmat friends are Onesmanam: "he who visits his friends only to look at their wives"; Asamanampo: "he who always looks backward while rowing"; and an old lady who, because of a single indiscretion in her youth, will forever be known as Aotsjimae: "the one who gets canoes dirty with her feet before the lime decoration is dry."

All of which is useful, but not yet enough. An Asmat can have a spirit name, a human name and a nickname, but still play no active part in the community unless he, or occasionally she, also has an *owom* name. This is the same as the human name of someone well known from another village. It cannot be borrowed or bought. It has to be taken, and it cannot be transferred until the one who owns it is dead. It is a headhunting name.

The ceremony which marks this transfer is truly beautiful. The initiate is seated in the prow of the giant ceremonial canoe and paddled off downstream by seven of the village's most influential elders. They travel through the swamps and out into the delta region, where the river mouth may be miles wide and fringed with thickets of mangrove. It is here, on the edge of Irian, that the shore is lined with bright white broken shell and a thin line of whispering casuarina. The canoe drifts with the tide across the mudflats and into the setting sun, keeping on going westward all through the night until it is twenty miles or more offshore, well out of sight of land.

At dawn the next day, they turn and face back into the rising sun. When the first light touches the canoe, the initiate is curled into a fetal position on the floor with his eyes tightly closed. He is being born again. Steadily the elders paddle back into the new day and, as the coast comes into view, the initiate's eyelids flutter. His hands open, he pulls himself up slowly into a crawling posture and when they enter the delta he is sufficiently well coordinated to cling to the edge and look out at the view with childish eyes. Once the mangrove gives way to river forest, he is crouched on all fours, and as the canoe slips under the canopy into a channel that forms part of the village territory, he moves to sit upright. Fifteen years of development are concentrated into the next hour or two, so that by the time the boat turns the final bend in the stream and comes into full view of the village, the transition is complete and a grown man stands tall in the bow, acknowledging the applause of everyone in the village who gather along the muddy banks, drumming, dancing and singing to welcome the arrival of a new warrior. And the name they use to greet him is his *owom* name, the name of a man who has had to die so that he might start to live. A name once worn with equal pride by the erstwhile owner of the skull the initiate carries tucked with reverence underneath his right arm.

What gives this rite of passage social and ecological substance is that the boy not only takes the name of the dead man, but becomes that man and assumes all his relationships as well. The victim's family will use the name when talking about or talking to the initiate, and welcome him as privileged mediator who can move safely in their midst. They call him father, brother or husband and extend to him the same care and concern they would display towards the man himself, had he lived. In truth, to them there is no ultimate difference. The dead man lives on, secure in the body of this stranger who has bored a small hole in the left temple of the skull, eaten his namesake's brain and taken on his soul.

The system has an integral and profound restraint. It requires that heads be taken only from among the ranks of those you know, at least by name and reputation. There is no point in having the head of a stranger without an *owom* name. So all Asmat is divided up into pairs of villages, usually neighboring villages who depend on the same hunting, fishing and food-gathering grounds. And these ecological twins form traditional rivalries, with each concentrating most of their headhunting activity on the other. Killings between them are still frequent despite the sanctions of the authorities in Java or the local missionaries, but they are anything but random and are never undertaken lightly, without good reason or proper warning.

The people in the two villages know each other so well that no raid ever comes as a complete surprise, and no battle ever continues beyond the one or two killings that ceremony requires. There is a ritual, almost balletic, rhythm to such encounters which involve a lot of formal posturing and the usual verbal preliminaries, but limit actual contact to a minimum. This is not war so much as the necessary and carefully controlled adjustment of local equilibrium. It is a game involving opponents rather than enemies. The rivalry can be intense, as it often is between those who compete for the same reasons and resources, but it is also polite. That may seem a strange word to use in the circumstances, but it fits. I have been in Asmat when equilibrium was disturbed unnaturally, by the accidental killing of a solitary fisherman mistaken in the undergrowth for a wild cassowary. The people on both sides were appalled. The situation between their two villages was stable and nobody wanted to be involved in an untimely spiral of revenge. There is no place in the Asmat system for meaningless feuds, so they found another solution—one drawn from a seemingly endless supply of rigorous social controls available to deal with almost any situation.

First, the man responsible for the killing was removed from the arena by a simple ploy. He was designated as "The Man Who Could Not Be Seen." He ceased to exist. He disappeared totally from social view, but was allowed before he vanished to nominate a friend to act as go-between. This "Man Who *Could* Be Seen" visited the murdered man's family on the killer's behalf. And on the agreed day, the victim's family came to receive their agreed compensation in the form of bows and beads, arrows, knives and blades directly from the killer's hands, thrust through the walls of the hut in which he was hidden. And where he remained hidden through all of a long day, handing out presents one by one, until all the family expressed themselves satisfied. Then he put a typically Asmat spin on the proceedings. He produced the most valuable present of all—an exquisite ritual axe-head carved entirely from a single piece of quartz crystal. This was clearly something in excess of the agreed sum, but the family recognized it for what it was, and took it and left, honor suitably restored. This parting gift, I discovered later, was an inspired device for finally healing the rift between the villages. Because the next day, "The Man Who Could Not Be Seen" reentered society, becoming completely visible, and restored the broken equilibrium altogether by going across to his victim's village—where he was hailed as the missing man himself—to get the crystal back.

The whole sequence of events in this remedial action, and in all those which lead to headhunting and initiation, is riddled with respect. Respect for rivals and for the environment, because the Asmat have come to elegant terms with both. Almost alone among the people of my acquaintance, they make ecological sense. They take their surroundings and each other as they find them and, instead of trying to bend the world to suit their needs, modify their own behavior to make the best of what is available. This is what makes them and us human. An ability to rise above the genetic imperatives and to modify and mold them to a different purpose. The big brains we have acquired in learning how to deal with each other can be used to challenge the genetic rules, and to play a different game. We are products of our natural history, but we are also free agents in the sense that we can reclassify ourselves and each other, creating distances between us or closing up the gaps, redefining natural boundaries and redrawing the areas of genetic interest.

The fact that humans are, on occasion, both aggressive and violent presents the Asmat with no problem and requires no heart-searching or remorse. They have stereotyped and ritualized such tendencies, allowing them full and satisfying play in headhunting, while at the same time resolving a pressing environmental problem. Without denying their nature, they seem to have found a way of using it to their own advantage. I have never been amongst people more consistently good-natured, or less prone to unregulated personal violence. By giving aggression a socially approved form, a license to practice, with rules and appropriate displays, they have hit on a very effective strategy. Another of the universals by which our species may be recognized is our love of such regulation and ritual. And given half a chance, as in headhunting or football, we are often quite content to minimize killing or aggression in favor of its alternative—the ritual display.

There is hope in this, but only if we follow the Asmat example and learn how to bend in favor of that which best allows equilibrium to be established. They have succeeded in developing a social system of astonishing intricacy and sophistication, one so sensitive to environmental pressures that it allows them to thrive in an area where we, with all our brute technology, must struggle to make a meager living.

The difference between us is best illustrated by something that happened during my last visit. The Crozier Mission at Agats was by then well established and quite substantial. There are stores and schools and sawmills on their riverfront, the only navigable one along that coast. A boardwalk between the landing stage and the various mission buildings allows the priests to ply their trade without getting their feet wet and, to one side of this intrusive complex, beyond the last tin roof, an area of tidal swamp has been cleared to form a football field. Here, every Saturday afternoon, teams from the nearest villages meet each other in a noisy, mud-splattered, soggy but nevertheless enthusiastic game of soccer.

This bush league has been going on now for some time and despite the intense nature of the competition and the fierce struggles which ensue for possession of the ball, every single game for almost a decade has ended in a draw. After a match that I watched, the referee—a recently arrived priest driven to the point of exasperation by his umpteenth game without a decision—remonstrated with the rival captains.

"Don't you see," he said, "the object of the game is to try to *beat* the other team. Someone has to win!"

The two men looked at him with compassion, reconciled to the fact that he was young and had a lot to learn. They shook their heads firmly and said: "No, Father. That's not the way of things. Not here in Asmat. If someone wins, then someone else has to lose—and that would never do."[408]

Of course not.

Such clarity is rare in human affairs. Much more often, there are factors which distort behavior in unfortunate ways. Some of these are postcontact distortions, imbalances of trade and territory created by colonial or other outside intrusion. There is good reason, for instance, to see occasional outbursts of random killing, which do occur in New Guinea today and amount almost to genocidal attacks, as largely artificial and economic. Sparked by competition for, and jealousy about, new trade goods such as horses and guns. But there is equally good evidence to suggest that traditional restrictions on aggression can break down for other, more natural, reasons.[42]

On the border between Venezuela and Brazil, the Yanomamö—about 10,000 of them—have earned themselves a reputation as "The Fierce People." The reputation is well deserved. For as long as any of their neighbors can remember, they have raided and murdered with apparent relish, killing, stealing, carrying off women as the spoils of tribal and intertribal conflicts that are frequent enough to be described as "chronic."[63]

The American anthropologist Napoleon Chagnon went to work amongst the Yanomamö in 1964, expecting to find that reports of their "brutal, cruel and treacherous" behavior were superficial judgments made by outsiders who had not got close enough to understand them. He was disappointed. Even after several field trips over a number of years, he was forced to admit that they had an unusually high capacity for rage and an unusually low threshold for violence. Chagnon's reports make it clear that hostility is endemic to Yanomamö men who assess each other largely on their capacity for violence and treachery, and on their willingness to fight. They practice a social politics of constant brinkmanship with each other and with all outsiders, poised always somewhere between bluff and explosive aggression. Conditions so reminiscent of the lives of young street warriors in many of our inner cities that it is tempting to look to the Yanomamö for clues to the origin of this destructive condition.

Chagnon discovered that violence and conflict among the Yanomamö were almost always about sex. When fighting took place between villages, it was initiated by the abduction of a woman, or in retaliation for such an abduction. At the end of every encounter women changed hands like pawns in a terrible chess game with live pieces that are valued but treated with contempt. When fighting broke out within a community, it was usually to do with sexual jealousy over infidelity or adultery. Women, it seems, are the currency of Yanomamö violence, made so by the peculiarities of a culture which has left them in short supply.

"Kinship" is one of the oldest and most revered tools of classical anthropology, producing lists and diagrams of perceived affinities that are tedious to read and often inaccurate, but they do at least reflect the true complexity of such matters in tribal social life. People in close communities have an astonishing awareness of genealogy, using proper names for relationships as distant and tangled as the cousin of one's second wife's younger brother. Such genetic sensitivity is vital to kin recognition and to the calculation of proper behavior. But among the Yanomamö it has become confused by the introduction of adjustments necessary to solve a problem of their own making. For a start, they practice female infanticide. This begins with a custom of killing any new child if the mother already has a nursing baby. That much is not unusual. Many human societies continue the biological practice of reducing competition among offspring by selectively starving or actually killing some of them. Sometimes this is necessary. But the Yanomamö skew their sex ratio in the process by showing a preference for male babies, killing female babies far more readily. And this inequity is compounded later by the grouping of adult men into defensive alliances or strategic lineages which artificially exclude some of the already scarce women by putting them into taboo categories which make them unavailable for marriage, and true blood relationships can be very hard to disentangle from a web of political expediency.[65]

Chagnon shows how this distortion of sex ratios has forced the Yanomamö to reclassify kin all the time, moving them up or down in generations, manipulating patterns to bring about the desired result, adding sexual factors to an already unwieldy and unnatural equation. And such confusion is compounded even further when women who, as blood sisters, nieces and aunts would normally be unavailable to any of their relatives for marriage, are transferred to categories of convenience in which they become accessible for reproductive purposes. And he shows how quickly and how often this artificiality can lead to ill-feeling and hostility—all directly due to a shortage of women.

In bioeconomic terms, when a resource becomes scarce, competition for it becomes rigorous and this usually leads to the development of new sources of supply. Where women are concerned, the best solution for the Yanomamö has been to avoid all the complications attached to those in the tangled tribal web, and to look for others, new women, without the burden of such association. So they habitually steal women from other tribes. They actually battle over women, and the most aggressive men naturally have the greatest reproductive success. Men who have killed have twice as many wives and three times as many children as men who have never killed. And this breeds more killers. Three-quarters of one large Yanomamö village that Chagnon examined turned out to be the direct descendants of one head man—truly "the father of his people" and the natural source of persistent aggression among his many male offspring.[64]

The spiral here is real and inevitable. As Yanomamö villages grow, tension increases to a point where it can only be resolved by fission, by breaking the community up into splinter groups that are more easily managed. But, in a world of relentless conflict, such groups are too small to feel secure and are compelled to resort to even more extreme expressions of hostility. Small wonder then that the Yanomamö, locked into a cycle of rapid population growth and range expansion, and committed to an endless round of violence, are "fierce." Who wouldn't be?

The only restraints built into the Yanomamö system are recognized grades of conflict which begin with bare-handed chest-pounding duels, progress to club-fighting competitions, and pass through spear-throwing encounters before escalating to outright open aggression. But restraint is not as attractive to the Yanomamö as it is to the Asmat, who reward and pay at least equal honor to the man who heals rather than kills. "The Man Who Can Be Seen" was hailed on his return from negotiating a crucial compromise with all the ceremony and respect accorded those coming home from a successful headhunt. His welcome put me in mind, briefly, of Neville Chamberlain's return from Munich in 1938—with none of the uneasy overtones of that short-lived appeasement. The Asmat are masters of long-term reconciliation with access to a vast repertoire of social bandages designed to deal with both individual and communal pain. They seem to know where and why it hurts and how to fix it. And, most important of all, they have a fine sense of what is appropriate and how far they can afford to grow.

Keith Otterbein, an evolutionary anthropologist, took a close look at the warlike behavior of forty-six cultures, ranging from Jivaro headhunters to feudal Japan, and concluded that humans are alike in that we fear and distrust strangers and tend to try to solve conflicts by aggression, usually a little more than the situation requires. We are also inclined to partition people into friends and enemies and to treat them well or badly as the diagnosis, and our genetic heritage, demands. And we are most likely to be warlike when our populations are out of balance. When we fall under the influence of the Second Pathic Principle as groups become too big to be personal and manageable, or as they become too small to feel secure. This much is given, but we do have other alternatives, some just as securely grounded in good, basic biology.[296]

If the Asmat can be described as fierce, it is only in their insistence on proper behavior. They have looked at their environment on the Casuarina Coast, assessed the abundance of sago palms which provide their main source of carbohydrate nutrition, and decided that human behavior has to be curtailed in the same way as palms are protected, by elaborate patterns of land use. They have laid down the boundaries and even adopted a palm sort of strategy for dealing with those who transgress or who need to make way. The remedy is a simple one—decapitation. And this apparently brutal solution has been softened and streamlined into an amazingly delicate and considerate system by one simple, vital and inspired ploy. The Asmat make sure that killing has direct personal consequences for the killer, who not only takes on the name of his victim, but also assumes that person's relationships and responsibilities. Killing, as a result, is something no man in Irian ever takes lightly. Murder is not just something you do to another person. It is something you do to yourself. It is a burden you accept only with considerable forethought and caution. In Irian, eating people isn't wrong, but no Asmat ever pretends that it is easy. The right things seldom are.

I find it very hard to judge the Asmat harshly. Killing makes me uncomfortable, but when I see what is done, and how and why, I am a little envious of their certainty and poise. They even handle envy with aplomb. If anyone admires or covets something that you own in Asmat, you give it to them. That puts the envious person under an immediate obligation to you which can only be discharged by giving you something, or rendering you a service, of at least equal value. Envy is expensive and, as a result, very rare in Irian.

So are sexual jealousy and adultery. It is strictly forbidden for anyone to have an affair with the wife of someone who is not both in the village and in total agreement. So, if you covet your neighbor's wife, you and he get together and agree to become *papisj* partners, to engage in a ritual exchange of wives. But *papisj* is very much more than wife-swapping. It has to be entered into with the full approval of both wives, and in the understanding that it will involve both men in lifetime obligations to come to each other's help in all time of trouble. And it puts a particularly heavy onus on the man who started all the trouble. He is honor-bound to ensure that the husband of the woman he covets never comes to harm in battle. So he knows, right from the start, that his lust can easily end up costing him his life.

The whole maneuver is inspired and so typical of Asmat in that it succeeds in turning what in other societies would normally be a divisive and destructive situation into a powerful force for social cohesion. Envy, greed and lust are not deadly sins in Asmat, just human foibles to be handled with the same sensitivity and skill the people show toward their whole environment. Gluttony is unknown in a community where generosity is the rule; sloth impossible if you want to stay alive; and wrath is taken care of by combining it with pride into a tradition of successful headhunting. By canceling out each of Christianity's seven deadly sins in this cavalier fashion, I am not pretending that everything in Irian is lovely. It isn't. Life, and adjustment to it, can be as hard there as anywhere else, but I am seriously suggesting that headhunting, the practice of ritual killing, is not necessarily evil. If it serves to promote ecological equilibrium, population stability and social cohesion—as it does in Asmat—it is well worth considering as a valid choice of lifestyle. Perhaps murder and even cannibalism are not bad in themselves. There are good arguments in their favor, as there are for abortion and the death penalty in our societies.

Everything depends, as usual, on context. So I want, in this chapter, to explore the circumstances in which we kill one another and to see if we can put these into biological and ethical perspective.

Some people never seem to kill each other at all. People such as the Senoi Semai of the Malay Peninsula, who are anything but fierce. Roughly 13,000 of them, representing the remnant of an ancient population, live in rainforest fractured by mountain and river. Like other forest people, they hunt, gather and practice simple shifting agriculture as appropriate to the season. But what sets them apart from most other humans is their insistence on being totally nonviolent.[106]

Semai are conditioned from childhood to believe that thunderstorms and elephants and strangers may be aggressive and dangerous, but Semai never are. "We never get angry," they say. Being human, of course they do get angry, feel hurt and hold grudges. But their belief in their own nonviolence surmounts such feelings and prevents Semai from putting each other under pressure as a result. They are not violent people, simply because they *believe* they are not violent people.[322]

This happy attitude allows Semai to meet force with flight. Running away from a confrontation is not cowardly to the Semai, who put no store in courage or bravery as virtues. Prudence in such matters is simply further confirmation of the character Semai have invented for themselves. This is a skillful and useful piece of social sleight of hand, and it works. Murder is not a recognized crime, nor is it an evil ever considered in Semai life. It is unthinkable and so it never happens. They have no need of any institutional way of preventing or punishing violence, no local police force or court system. And no single instance of murder, attempted murder or even injury by a Semai has ever come to the attention of government or hospital authorities in Malaysia.

So Semai meet force with passivity. Violence terrifies them and they do everything possible to avoid it, caring nothing that others may call them cowardly, taking "timid" as the ultimate compliment. And it seems that what makes this possible is rigid internal self-control and cultivated restraint, because when conscripted into the British Army and sent to deal with Communist insurgents during the peninsular crises of the early 1950s, Semai were swept up in a kind of insanity they called "blood drunkenness."[323]

"We killed, killed, killed," they said later. "The Malays would stop and go through people's pockets and take their watches and money. We did not think of watches or money. We thought only of killing. *Wah,* truly we were drunk with blood."[106] One even told of drinking the blood of a man he had just killed.

"Out of character" behavior of this sort is not uncommon. The Avatip, who live on the Sepik River in New Guinea, also see themselves as essentially peaceful. A description belied by their long history of intercommunal conflict. They fight as often as their neighbors, and as hard, but insist that this behavior has nothing to do with aggression. They do not feel anger or hostility toward their opponents, they claim, but talk instead of putting on a different or "bad" face. An alter ego made visible in the form of a chest ornament in the shape of a face, or in a terrifying mask. And they change their personalities in the process, becoming "deaf" to their enemies, putting themselves beyond the reach of sympathy, remorse or shame. They manage this all by magic, by becoming possessed with the spirit of an ancestor or a guardian who relieves them of responsibility for killing.

It is a familiar device. Returned to their communities after the war, the Malaysian Semai were as gentle and noncompetitive as ever. Their violent outbursts were quickly forgotten, set aside as something done by another person in another country, so completely dissociated that they were able to keep their nonviolent image of themselves intact. This powerful sense of self, however contrived, is clearly useful and seems to have been cultivated by a few simple cultural devices. First, the Semai never punish or condemn aggression in their own children. They simply fail to give it any currency as a possible alternative. They refuse to take it seriously, laughing at children who show aggression and try to hit each other or adults. And secondly, they deny children the most common expression of social aggression. They remove all possibility of rebellion by allowing the practice of *bood.* This translates roughly as: "I don't feel like doing it," but what it means in practice and in functional terms is that no adult Semai can make a child do anything it doesn't want to do. If the child chooses to *bood,* the matter is closed. To bring any pressure to bear on the child would be wrong. That never happens, and more importantly, no Semai child has cause for the kind of resentment that breeds rebellion and provides a convenient excuse for acts of displaced aggression or delinquency, either then or later in life.

The Semai system sounds inspired, almost too good to be true. A social solution to rival the Asmat one. On Semai terms, in their environment, amongst themselves, it certainly seems to work wonderfully well. But where it breaks down is when Semai are forced to deal with outsiders who think and behave differently. There is no hierarchy or system of authority in Semai society. Ordering others to do things is impolite, ordering them to do things they don't want to do, impossible. So there are no chiefs, no headmen and no way of organizing Semai society to let them deal profitably with outsiders. They are ill-equipped for competition and the open market. It is no place, anyway, for gentle people.

Semai life, on the other hand, is very seductive to some outsiders. The lack of violence, aggression and bureaucracy, the abundance of kindness, generosity and good humor led early British anthropologists who came to work among them to settle down to live there with attractive Semai wives. They, and more recently Robert Dentan, became extremely fond of their subjects, which is not always easy for anthropologists to do, even after getting over the disillusion that comes with awareness that the "noble savage" is seldom more than a myth. But the Semai are indeed touched with nobility and have succeeded admirably in finding ways of handling human violence. They do have something important to say to the rest of us about good and evil.

The Semai, like many other tribal people, have a fine notion of "natural order." They believe that there is a way things should be. To describe it they use the Malay word *patud* which means "fitting, proper, fair and right," which is precisely the sense I have been suggesting lies behind the action of the Goldilocks Effect and informs the kind of moral philosophy Aristotle described as "just enough." Not too much or too little, but just enough of the right stuff, the kind of stuff that makes for good ecology.

The Semai judge everything on the basis of whether or not it conforms to this ideal. For them, aggression is not *patud,* but then neither is discipline nor pressure brought to bear to prevent aggressive acts in others. Things have to find their own proper level. A lack of generosity with food is not *patud,* but neither is a tendency to ask too often for such food to be shared. Physical or mental disabilities are not *patud,* but neither do they become a source of concern unless they make others unhappy. Noninterference in the affairs of others is an essential part of the Semai ethic of nonviolence—and both of these overtly negative qualities are regarded as *patud,* as are withdrawal and passivity in the face of interference or aggression. Infidelity and illegitimacy are not *patud,* but neither are necessarily bad nor condemned as evil. If the love affair is short-lived, the Semai call it "just a loan"; in the same way that the non-judgmental Seychellois refer to children born out of wedlock simply and happily as "spares."

In summary, what may most usefully be said of the Semai is that they do not deny human nature, but do their very best to hobble it wherever necessary, in the interest of balance and human welfare. Their utopian solution, however, is a poor competitor in the open market. And as all ideas, all mutations, in the end all species, must be judged on their success in competing with others of their kind, the Semai experiment has to be judged a noble failure.

The Asmat fare better in the open market. They give as good as they get and are less squeamish about admitting that, as a species, we are predisposed to respond with aggressiveness to sources of disturbance. They are also a lot more realistic in designing a response that takes advantage of our tendency to be competitive, and in using it to enhance inclusive fitness. In sociobiological terms they must be deemed a success, even if in the end they fall prey to ideologies supported by larger numbers and superior weaponry. I see them, at the worst, as heroic failures.

While, of these three essentially aboriginal cultures, the Yanomamö are easily the least successful in evolutionary terms, letting themselves become so confused by what Claude Lévi-Strauss has described as "cultural *bricolage*" that they suffer from runaway aggression and real overkill. They are living examples of what can happen when the pathic principles are allowed free play and produce something that can only be described as pathological. They offer a suitable warning of the dangers we face in letting any of our genetic imperatives—in their case it happens to be sex—run away with the rest of our lives, overwhelming our natural sense of order and fair play.

I believe such a sense exists and that it arises out of acts of reciprocal altruism, out of the kind of sharing that is common in hunting and gathering societies, but has been suppressed by the acquisitiveness that began when we settled down and accumulated surpluses large enough to make some among us richer than the others.

I find it useful to look back to our roots and to try, whenever possible, to meet people such as the Asmat who still live in much the same way as our species has for over ninety-nine percent of human history. Any strategy that lasts for several million years is worth taking seriously. It has earned its evolutionary spurs, done what it takes to be described as a success in Darwinian terms. While "civilization," the lifestyle that we have been experimenting with during the last 5,000 years—just 0.25 percent of our time in the arena—has to be described as an interesting, troubled and still unproven solution.

The problem with history is that, like tabloid newspapers, it tends to concentrate on the lurid bits. In the words of Thomas Hardy: "War makes rattling good history; but peace is poor reading."[175] The fierce Yanomamö keep making the headlines, while few have heard of the Semai, and even fewer of the Xinguanos, a collection of peaceful people who, despite their different origins, have succeeded in keeping intertribal conflict to a minimum in the Upper Xingu area of the Amazon.

There are ten villages there, with a total of perhaps 1,500 people, speaking four major languages. They live in a basin that lies at the foot of the high Mato Grosso, where the Xingu River has its sources, running clear out of the old continental bedrock, not muddy like the main Amazon, and free of most of its troublesome insect pests. Life is relatively easy there, fish are plentiful and until recently there was little contact with the outside world. Things are changing, of course, but the ten autonomous communities, though often suspicious of each other, continue to have a strong sense of their own unique identities. All they have in common is an antipathy to violence, and customs which discourage aggression among their children and reward peacefulness among adults. Individually, they are not unlike the Semai, but they seem to have gone one step better and found a way of dealing effectively with other people as well.

The ten villages, who may each have sought refuge in the Xingu from violence elsewhere, have succeeded in building a community of nations in miniature, enjoying the advantages of trade, intermarriage and intertribal contact that saves them from the fragile isolation of the more introvert Semai. And though their numbers remain small, the people of the Xingu have been able to carry this community spirit over into their growing relationship with the Brazilian government, expressing their collective concern in ways that seem likely to win respect and autonomy for the Xingu Basin.[160]

What the ten tribes share is a belief that what is "good" is tied to peacefulness. A good man avoids confrontations and rarely shows anger. Anger disturbs and frightens all Xinguanos, who see it as something like a wildfire over which one has no control. A good man exercises necessary constraint and makes a good citizen because he responds to the feelings of others. He has empathy and a capacity to share their pleasure and their pain. In the language of one of the tribes, the Mehinaku, he has *ketepepai*—a quality sometimes also seen in well-made objects, things that have beauty and balance.

The Xinguanos carry their feelings about peacefulness over into their system of intertribal trade. Everyone in the community values shell belts and necklaces, stone axes, salt, cotton, hardwood bows and ceramic pots. They all measure their wealth in these items and each has the resources and the knowledge necessary to produce any of them. But they have, over the centuries, contrived a system of sanctions and customs which deliberately protect monopolies. Everyone knows that only the Caribs make shell ornaments; no one produces or trades in salt for fear of upsetting the Mehinaku; and nobody would want to use a ceramic unless it was made by one of the much admired Waura women potters. Such conscious controls are good for trade, and trade means trust, which provides a relationship which is valuable in its own right. One that is equally deliberately fostered by encouraging intertribal marriages.

The other major point of agreement amongst the tribes is that authority is something to be regarded with a certain suspicion. Each village has a spokesman, an elder who represents them in trade and who plays a central role in ceremony and tradition, but the situation is best summed up by an anthropologist who said: "One word from the chief and everyone does exactly what they please." And what pleases them most of all is singing, dancing, trading and wrestling. The last activity being one of the few diversions Xinguanos need to redirect or sublimate aggression—their substitute for conflict. "When we are angry," they say, "we wrestle; and then the anger disappears."[160] In fact, they wrestle very well.

The most interesting aspect of the Xingu solution is its rarity. I know of no other situation anywhere in the world that allows peoples from different backgrounds, speaking totally different languages, to coexist in such harmony. Conflict is the rule elsewhere. National, ethnic, racial and religious divisions everywhere else seem to lead inevitably to conflict. Most social and political mixes are volatile, the genes are obliged to be unkind to outsiders, and aggression is so easily aroused and so contagious once it is released that intergroup problems are epidemic.

The recorded history of eleven European countries during the last 1,025 years shows that they were engaged on average in some kind of military action forty-seven percent of the time, or about one year in every two. The lowest scorer has been Germany with twenty-eight percent, and the highest Spain with a massive sixty-seven percent, waging war in two out of every three years throughout the last millennium.[361] One study covering just the twentieth century shows that an average of three high-fatality struggles have been in action somewhere in the world at any moment since 1900, leading that researcher to the sad conclusion that such conflict is "a routine, typical, and thus in fact normal, human activity."[410]

Research has concentrated in recent years on the physiological and neurological source of so much aggression. Controversy still surrounds most of the results, but some simple conclusions seem warranted.

Most of our agonistic behavior appears to have its origins in an area lying low in the central brain. This is known as the *limbic* system, from the French for borderline—"on the hem." There are three groups of cells here which, in mammalian brains at least, appear to produce marked changes in behavior. Damage to them reduces aggression and fear, effectively eliminating the genetic tendency to distrust and to be nasty to outsiders. Stimulation of them, particularly of the cells of the hypothalamus, increases hostility, suggesting that this may be the "seat of aggression," a possible site for the switch the genes have arranged to further their selfish needs. It appears to be a simple on/off switch, the kind made possible by a single mutation, producing arousal of a generally aggressive kind. The direction and control of such feelings is almost certainly arranged at other levels of the brain, such as the frontal lobes, where action can be fine-tuned to focus on particular objects of aggression, such as those individuals lower in the hierarchy, or those simply labeled or recognized as social or genetic outsiders.[181]

The frontal lobes, of course, can also be responsible for the inhibition and conscious control of general aggression. There is clear evidence that damage to this area in man can cause "uncontrollable outbursts of explosive aggression," actions without any evidence of remorse or concern. Starting in 1848—when an explosion sent an iron rod rocketing through the head of Phineas Gage, destroying large areas of his prefrontal brain, turning a mild-mannered man into an erratic, aggressive person no longer able to hold down a job or deal effectively with his fellows—it has been recognized that "lower" and "higher" parts of our brains need each other and work together to channel and direct chemical transmitters and hormones.[95] Radical experimental techniques such as prefrontal lobotomy—a procedure about as subtle as shoving an iron rod through someone's skull—have succeeded in interrupting such flow and turning patients into manageable vegetables. But the details of control remain mysterious.

Serotonin, a chemical that has a stimulating effect on the nervous system in high concentrations, could be involved; but no trigger has yet been discovered to induce its flow. Male hormones, such as testosterone, heighten masculinity and the aggressiveness that appears to be associated with it; but history is replete with examples of violent eunuchs, such as the Byzantine general Narses, who destroyed Vandal power on behalf of his emperor Justinian in the sixth century. All we can be sure of is that fear and rage have a neural origin, probably in the lower part of the brain, which is stimulated by threats perceived and identified in the higher parts of the brain, and that these two areas are in constant chemical contact.

Genetic influence, involving a predisposition toward greater or lesser violence in certain circumstances, seems also to play a part. In 1978 a nameless Dutch family were discovered, some of whom were unusually prone to impulsive outbreaks of aggression, such as running over employers when reprimanded for poor work, setting fire to buildings for no apparent reason, or attacking people with pitchforks and knives. Fourteen individuals were involved and urine samples from these showed that all were deficient in an enzyme which usually controls neurotransmitters such as serotonin. All were also male; there was no record of such behavior among the female members of the family. This suggested a recessive genetic factor lying on one of the X chromosomes, the one that all men inherit from their mothers. Daughters get two such chromosomes, one with the mutation in question, and one with a good copy of the same gene that masks its bad effects. After a search lasting more than a decade, a team from the University Hospital at Nijmegen has finally located this gene, whose action does allow an accumulation of chemicals that make it difficult to deal with stressful situations. This is not necessarily "the aggression gene," elimination of which will bring about universal harmony and brotherly love, but it is evidence at least that some aggression, in some people, under some circumstances, is indeed under genetic control.[48] There will almost certainly be more such discoveries.

There are, however, cautionary tales aplenty in this kind of research. Adrian Raine, a psychologist at the University of Southern California, thought he was on to something recently when he did brain scans of forty-four long-term prisoners and discovered that just one of them, a serial killer responsible for forty-three deaths in five years, had a peculiar blob of activity in the area of the thalamus. Because he needed a noncriminal control for the study, Raine had his own brain imaged and found the very same blob there, though he swears that his only link with crime is the research he does into genetic links with criminal behavior.[270]

The search goes on, but as of now science still cannot predict when any individual will display violence or explain exactly why some groups of individuals will combine to fight others. All we know for sure is that our species can be aggressive, sometimes unreasonably so. Freud attributed this tendency to frustration of the sexual drive by the ego, and described aggression as an inborn urge that periodically breaks through the dams of inhibition.[134] Konrad Lorenz, the father of ethology, modernized this Freudian view with a mass of new data from animal behavior.[245] He concluded that aggression is innate, a general instinct we share with most other species, that usually depends upon specific releasing stimuli, but can explode spontaneously unless provided with appropriate outlets. And the frustration lobby seems to have gathered steam in psychology and psychiatry with a widespread assumption that, while conditioning and learning might inhibit aggression, the drive itself is largely predetermined and probably not amenable to more than superficial social reform.[34]

Such pessimism seems ill-founded, if only because it is not possible to talk about "aggression" as a unitary behavior pattern. It consists in practice of a wide array of different responses—some territorial, some hierarchical, some sexual, some defensive, some disciplinary and only a small proportion directly hostile. And all of these are capable of distortion under pathic conditions in ways that could be considered antisocial, perhaps even evil. Most studies of so-called aggression in humans equate it with violent and hostile behavior intended to hurt other humans, and tend to concentrate on the conflict such actions cause.

I find it more than a little ironic that the study I quoted earlier, which concluded that warlike behavior was a normal human activity, was published in the *Journal of Peace Research,* which is regarded as the foremost interdisciplinary periodical devoted to peace studies. But a quick glimpse at any issue of that journal makes it quite clear that it is dedicated instead to studies of conflict and the causes of war. The editor himself notes, with some sadness, that of over 400 articles published by his journal in twenty years, only one was devoted to "the empirical study of peaceful societies with a view to finding out what seemed to make them peaceful."[415]

The reason is simple enough. Peaceful societies are very hard to find. One survey of 130 social systems found just six that could be described as peaceful in the sense that they were not involved in collective violence and provided no special role for warriors.[357] And at least one of these six ought to be excluded on the grounds that feudlike killings sometimes take place. The numbers involved may be small, but for a population of "peaceful" !Kung Bushmen in which twenty-two such deaths have been reported, this is a high percentage, as high as any recorded for homicides in our "violent" societies. So if the criteria for peacefulness are expanded to include a lack of interpersonal violence and the possession of sanctions which preclude such violence, the list becomes very short indeed. It is limited then to those who live on islands or mountaintops, on Arctic wastes or in isolated areas of dense rainforest, where people have been able to evade the aggressiveness which seems to be epidemic everywhere else.

Conflict is easy to understand. It is peace that needs explaining. The Semai keep theirs by using the forests and mountains of Malaysia as natural barriers, seeking solace in the common observation that translates best as "Good fences make good neighbors." The Xinguanos understand that, but have put a number of well-designed gates in their fences, which not only allow intertribal contact, but encourage assimilation. Their most impressive trick has been to keep the peace even while engaging in rich and close intercommunal relations, despite the fact that interpersonal violence can be found at its most extreme between those who are most closely related. Minor problems in such situations have a way of exploding elsewhere into major confrontations, but not so far on the Xingu River.

Why? It is hard to be certain, but the impression I get from being there is that they have found a very narrow and delicate balance. A point of precious and probably limited stability midway between "positive peace," that depends on the openness necessary to exchange goods and people; and "negative peace," that arises from avoidance and the perception of others as different and probably a little dangerous.

We all need the guidelines provided by those who represent or embody the things we identify as "bad" and try to avoid. The Xingu tribes have discovered that good is meaningless and very difficult to sustain in the complete absence of evil. So, while championing nonviolence and generosity, they have also created its antithesis. Despite the lack of any real evidence that such people or such practices truly exist, they encourage talk of witches and witchcraft, of vengeful beings who hold grudges and nurse imagined slights. The punishment for witchcraft is death, but in the interest of long-term stability, it seems the community is prepared to tolerate the sacrifice of an occasional individual for the cause.[160]

Sacrifice is one of the cornerstones of human history. It lies at the heart of so many religions that classics scholar Walter Burkert believes it may even have given human society its form. He suggests renaming our species *Homo necans*—"man the killer."[53]

There is something in this. A man shot to death by arrows lies buried at the main entrance of Stonehenge. And just a few miles away, in the center of the Bronze Age circle known as Woodhenge, archaeologists found the body of a three-year-old girl with a split skull.[54] The Greek historian Pausanias tells of the dismemberment and communal eating of a child sacrificed in the sanctuary of Zeus on top of Mount Lykaion—and raises for the first time the possible consequences of portraying the gods in human form.[298] Bridges, houses, temples and forts throughout Europe, Asia and the Pacific contain "foundation sacrifices" designed, it seems, to ensure that they were imbued with the proper spirit.[132] The Judeo-Christian tradition began with the sacrifice of Abel by his brother Cain, the aborted sacrifice of Isaac by his father Abraham, and the death of the son of God himself at Golgotha.[249]

Human sacrifice appears to have been the engine too of New World cultural evolution. The classic Mayans had a sophisticated ritual grammar, replete with verbs that describe decapitation and the tearing out of a still-beating heart. Blood was the mortar of their ritual life and human blood the most noble of all possible offerings.[336] But it was the Aztecs who carried this grim reality to its bloody zenith in the "Flower Battle," designed specifically for the taking of high-caste captives suitable for sacrificial death. Each of these chosen ones stood on a killing stone, armed with a sword of feathers against four warriors whose task it was to "stripe" him with their razor-sharp blades until loss of blood weakened him enough for one of them to flay him alive and wear his skin to a feast of the dead man's flesh.[72] And the Incas set up altars on the very top of the highest Andean peaks, where they buried beautiful royal children alive.[378]

Anthropologist René Girard describes sacrifice as "the most crucial and fundamental of rites . . . and also one of the most commonplace."[145] It is certainly still happening. Following a tidal wave in southern Chile in 1960 a five-year-old boy was thrown into the sea by Mapuche Indians intent on appeasing the ocean spirit. In 1986 an Aymara Indian man in Peru was decapitated in a revival of the ancient ceremony of "paying the earth" by cocaine traders hoping to bring fortune to a new venture. And later that same year evangelical Christians in an Andean village drove a stake through the heart of a nine-year-old boy in an attempt to save the life of a sick man.[378]

Ritual killing of this kind is hard to deal with. The Danish philosopher Søren Kierkegaard examined Abraham's willingness to sacrifice his only son for his belief and found it impossible to produce any logical justification for such a "monstrous paradox." The end simply doesn't justify a means so horrible to contemplate. So, Kierkegaard concluded, something else must be going on. Something "which no thought can grasp because faith begins precisely where thinking leaves off."[228] He was right. What we are seeing in sacrifice is the outcome of a battle between the genes and a newer, less predictable, human program. Which may be why

Aeschylus, Aristotle, Ionesco and Brecht were all fascinated enough to weave whole dramas around sacrificial themes; and why two of Steven Spielberg's most successful films—which must provide some sort of mirror for our times—have dealt with cults of human sacrifice.[98]

Psychiatry suggests that guilt plays an important part in such obsession, that we overcome fear and atone for our sins, particularly during an emergency, by seeking a scapegoat and offering this up in our place. Kikuyu in Kenya once allowed a man accused of incest to save his life by passing the guilt ceremonially to a goat that died in his place. Wealthy Moors in Morocco each kept a wild boar in their stables as a similar repository for all the evil in the home.[133] The nearest biology can come to this is in acts of appeasement and submission to a superior, which can become desperate if the social distance is too large to bridge by normal means. At a simple level, some lizards break off and abandon their tails in an attempt to distract a predator long enough to make good their escape. The Yakuza gang member in Japan who admits responsibility for failure saves his life by ritually sacrificing a part of his body in front of his superiors. He cuts off a finger. This makes genetic sense, but it is far more difficult to account for Agamemnon's sacrifice of his whole daughter Iphigenia in return just for a wind good enough to carry the Greek fleet to Troy.

The genetic inclination is to do no such thing. The arithmetic is against it. Losing a packet of half of your genes is something most parents will fight to avoid. Historical evidence suggests nevertheless that many of our most important sacrifices have been made precisely Abraham-style, by or with the complicity of parents, perhaps because no other choice could be so potent. If atonement is what you need, then maybe nothing will do but the sacrifice of the most precious thing you have—your only child in return for expiation of your sins.

In 1982 UNESCO investigated a child, an eight-year-old boy found freeze-dried in a tomb 17,716 feet high on top of Mount Plomo in the Chilean Andes. He was wearing a black llama-wool tunic and embroidered moccasins. His plaited shoulder-length hair was held in place by a headband, and his woolen cap was crowned with condor feathers. He bore a silver armband and a silver pectoral shield, both signs of noble birth, and was surrounded by regal gold grave goods. He was found with a peaceful expression on his face, as though he had just fallen asleep in a tumulus beneath five feet of earth and stone. Which is exactly what happened because, despite all the evidence of care and concern, the child was buried alive and froze to death before he woke five centuries ago.[392]

Was Abraham evil? Kierkegaard couldn't decide. And I am no closer to bringing in a verdict on the father of the Inca child. To both parents, the actions required of them were painful, but may well have seemed inevitable, mitigated perhaps by their belief that the child itself would be transmogrified in the process. Judgment can only be made, as with headhunting, in context. But I have to say that, like Kierkegaard, my mind and the program involved in this issue grind to a halt at the point where it becomes obvious that the end, however noble, cannot justify the means. There has to be some other solution that brings honor into the equation, this time on the side of the genes.

There is at least one modern example of a deliberate sacrifice along the same lines made by the terrorist from an extremist group, whose political affinity I forget. It doesn't much matter except that he believed in the particular "holy war" in which he was involved, at least enough to build a powerful bomb into a case he allowed his common-law wife to carry innocently aboard an aircraft, despite the fact that she was pregnant with his child. The genetic sacrifice is the same—half of his own genes—but that is not all. The equation is hopelessly unbalanced by the lives of all the other passengers on the disaster-bound airplane. That alone distinguishes this case from the Inca one and condemns the act as immoral and the actor as evil. It and he contribute nothing to the stability of their ecology, but are plainly inimical to it.

However, there is no war and no peace without a price. The Inca with his altar, the terrorist with his bomb, and the Xinguanos who feel that they need a witch-hunt to keep their fragile accord intact, all do their own arithmetic. And because they are human and know that the genetic rules are at least partly obsolete, they also know that there are other options open to them.

There are less draconian forms of action available for appropriate aggression. And less drastic ways of redirecting undesirable aggression. The Hopi Indians of Arizona are usually described as "peaceful." They do indeed suppress physical forms of aggression and even find wrestling undignified and tasteless. They are trained from childhood to smile at enemies and to share willingly, but they retain one very effective form of aggressive expression. Hopi are highly skilled in the devastating art of insult, attacking each other and outsiders with "tongues as sharp as poison darts."[113]

And if such cruel words are set to music, and given the ritual authority of melody, they seem to cut even deeper. In Greenland, Alaska and on the Aleutian Islands, Inuit people are compelled to spend long periods in the dark days of winter in very close contact with one another. The kind of concentration that has led to outbreaks, often lethal, of "polar stress" among those less well prepared for this aspect of arctic life. The locals know all too well how quickly anger can blossom in confinement and they settle all potential disputes by song duels, often accompanied by provocative dances and rude gestures. The lyrics are cruel, filled with "little sharp words like splinters hacked off with an ax," and those judged to be the most painful and inventive are greeted with applause that can hand victory even to the one who started out in the wrong.[188]

Being clever in the art of conflict resolution is clearly more important than simply being right. And if you can't be right or clever, it is well worth finding some other way of letting off steam. Central Australian aboriginal women take turns hitting each other over the head. In Baffinland, they boxed each other's ears. While German students fought carefully controlled duels designed to do no more than leave a ritual scar that became a permanent symbol of courage and pride. Having one often meant that you need never fight again. There is method in such madness, particularly if it can intercede in the process of escalation which blows disputes up into conflicts, and lets conflicts explode into full-scale wars.[34]

War, despite what the Prussian general Carl von Clausewitz might have believed, is *not* just a continuation of policy by other means. War represents a major change in human history, a shift from hostility to something far more organized and more lethal. A venture beyond what American anthropologist and cavalryman Harry Turney-High has called "the military horizon."[391]

In 1949, in the wake of the Second World War, Turney-High took his fellow anthropologists to task for ignoring a major facet of human behavior. He accused them of asking how people eat, drink, play, mate, court and count kin, but not how they fight. In *Primitive War,* he described how combat actually took place, analyzing headhunting, captive torture and ritual evisceration; examining the effects of clubs, spears and swords on human flesh. Making it quite clear that primitive man already had blood on his hands, but also pointing out that none of this mattered when it came to looking for the origins of our current military and political systems. The state, he suggested, only emerged when society moved from primitive conflict to something quite different, to what he called "true" or "modern" war. The transition took place, he concluded, with the appearance of an army under the orders of appointed officers. Then, and only then, was it possible for a state to come into being and to take decisions about its future and its policies—to decide whether it wanted to be aristocratic or democratic, theocratic or monarchic.

There is a simple functional difference between Yanomamö skirmishes and Zulu war, between Asmat raids and campaigns of the Austro-Hungarian army. The Yanomamö and the Asmat are interested in women and food, in simple resources; and their conflicts, while undoubtedly bloody, are strictly and severely curtained by ritual and custom. The Zulus, however, stood at the turning point and were carried across it almost singlehanded by their chief Shaka who, at the turn of the nineteenth century, cast all the old restraints aside. He formed regiments of permanent warriors who lived, not at home, but in military barracks under appointed officers, denied marriage until their fortieth year. He designed a new weapon, a stabbing metal-tipped spear with which he trained his men to disembowel, not just wound, their enemy. He taught them fighting drills, and he instructed them to kill, not just male rivals, but—like a lion moving in on another's pride—all the children of an enemy's ruling families. In short and in fact, this extraordinary man practiced true, modern war of the kind Clausewitz would recognize and endorse; a form of conflict that does not just count coup, but leads to victory, territorial conquest and the disarming and total destruction of the enemy.[295]

By 1967 Turney-High's distinction between "primitive" and "modern" war was finally recognized, and a new generation of young anthropologists set out to answer his question: "How does this group fight?"—and we finally have some insight into how war might have begun.

At one extreme are those like the Yanomamö who, with prudence and despite their fierce reputation, have settled for something short of true war. They indulge in ritual clashes which are never allowed to become decisive battles. For people like the Maring of central New Guinea, the struggles are more symbolic yet, limited to grand face-offs with neighboring tribes only once a decade or so, by the strict requirement of having enough surplus pigs to provide suitable sacrifices as thank-offerings for the survivors. They limit their ritual displays of strength to times when these also make ecological sense.[396]

The Maoris were something else. For a start, they were new to their area, arriving in New Zealand barely 1,000 years ago. An immigration which, even lacking all other evidence, we can be certain of because they made eighteen species of local bird almost immediately extinct. They were also well organized, the Zulus of the South Pacific, displacing weaker people, warring among themselves over grievances, learning that insult was unforgivable, storing up memories of such slights over generations, mounting an enemy's head on a stake where it could be symbolically degraded. This cycle was normal for primitive conflict, but the Maoris were also skilled in tactics, doing battle in the "modern" unrestrained way, aiming to kill rather than just intimidate. They had crossed the military horizon and might well have exterminated themselves were it not for one further fact. They built refuges in the form of forts, at least 4,000 of which have been found, and the attack and the defense of these, the sieges they entailed, slowed war down to a vital degree.

The Stone Age had formidable weapons—clubs, maces, spears,

daggers, slings and bows—which were used even in the most stylized ritual conflicts and continued to be used until the invention of gunpowder. The shift from primitive to modern war, however, was marked not so much by changes in these, but by the organization and the aims of those taking part. The British military historian John Keegan, in his brilliant analysis of the evolution of war, suggests that one of the major shifts took place with the construction of the first true *strongholds*—places not just of safety from attack, but of active defense. Places secure enough to provide stores of food and water, and at the same time strong enough to impose military control over the areas, the "killing grounds," they commanded.[225]

Such bases of action, Keegan points out, were a vital concept. The Romans made them almost portable, training their legions to throw one up at the end of each day's march into hostile territory. But too many forts or strongholds were a sure sign of weakness in military ecology. They marked an absence of central authority. All the great states defended themselves at their periphery, on the frontier—hence Hadrian's Wall, Offa's Dyke and the Great Wall of China. Within such barriers, there could be safe roads and open cities, all the advantages of the *Pax Romana;* but where local redoubts, forts and castles proliferated, these led inevitably to endemic local conflicts and the breakdown of statehood.

Keegan also makes a nice distinction between agriculture and the pastoral way of life, drawing attention to the difference in diet and mind-set these styles produced. Pastoralists had the military advantages of polished cooperative hunting routines, a familiarity with the techniques of slaughter, and an unsentimental attitude towards their flocks. "They knew how to break up a flock into manageable sections, how to cut off a line of retreat by circling to a flank, how to compress scattered beasts into a compact mass, how to isolate flock-leaders, how to dominate superior numbers by threat and menace, how to kill the chosen few while leaving the mass inert and subject to control." They were, in short, perfectly prepared for most of the rigors of modern war.[225]

Such talents ushered in the age of the war horse and the fighting chariot, the time of the Hyksos, the Hurrians and Kassites; the

Assyrians who "came down like the wolf on the fold"; and the Huns, Turks and Tartars.[60] The speed and agility, the skill at arms of these mobile warriors held the Middle East in thrall for centuries that culminated in the formation of the great Persian empire. And this only fell when confronted with an even more military development that took them completely by surprise.

This marked the return of the farmer and began in Greece, whose mountainous terrain made horses inappropriate. And it involved rugged footsoldiers who were also farmers of small holdings they could not afford to leave for too long. They favored settling matters as quickly and decisively as possible. So it was through and for them that the Greek *phalanx*—a tightly massed rank that fought in a very narrow field—came into being, producing a terrifying, short-lived clash of bodies and weapons at the closest possible range. This, for the first time ever, was war without quarter, battle without ceremony, in which flags and uniforms and elegant maneuvers counted for nothing. It was cutthroat and back-stab conflict, hand to hand, hard and short, with a swift follow-up designed to dispatch the wounded and to lay waste to the land.[173]

Such fighting, face to face, with death-dealing weapons and without restraint, defies nature. There was no precedent for it in human or natural history, and no defense possible in Persian ranks accustomed to more procrastinated conflict, in which ritual and restraint were as common and as time-consuming as actual combat. So they turned and ran. And the young, audacious Alexander became Great, defeating Darius in three quick and decisive battles despite overwhelming odds, bringing the empire to an end and changing the face and the character of war for ever.

There have been other substantial developments since then. The invention and use of gunpowder, the expansion of combat to the sea and the air, sophisticated logistics and supply—and these have all turned war into a major industry, the largest single item in most nations' budgets. But, with the possible exception of nuclear weapons, which may make large-scale conflict difficult to comprehend or allow, nothing has been added to the nature of war since the Greeks made it such a nightmarish occasion. That turned war from a glorious undertaking to something General Sherman dismissed at the end of the American civil conflict with the bitter words: "I am sick and tired of war. Its glory is all moonshine. War is hell."[354]

To overcome the notion of personal aggression without the option of submission, of lethal risk without the possibility of appeasement, modern war has had to pretend to be something other than what it is. But we are running out of excuses. Technology so devastating that it threatens the survival of the planet has put paid to pretense that war is a continuation of politics by other means. The Roman evasion: "If you wish for peace, prepare for war," once seemed a valid argument in favor of deterrence; but it looks threadbare in the face of armaments that seem to guarantee nothing but mutually assured destruction. And the notion of a good or "just" war becomes difficult or impossible to justify when both participants claim to have the same god on their side.

Saint Augustine of Hippo, in the fifth century, made what remains the most worthy attempt to live with inevitable war by saying that taking part was permissible, not sinful, as long as it met three criteria: the cause must be just; the action must take place under proper authority; and it had to be waged with "right intention"—by which he meant to achieve good or to avert evil. But no one has been allowed to make his or her decision on anything like these ideal terms since Machiavellian politics gave the state all the justification it needed to do whatever it pleased. "It's in the national interest," they say. "Line up over there and we'll bomb them back into the Stone Age." Back below the military threshold.

Actually that wouldn't be a bad idea—at least the bit about the threshold, because back then things were a lot more simple. They have become more complex, but the direction things have taken, the pattern of the evolution of war, was pretty much inevitable. It followed the principle of Darwinian natural selection. Our early ancestors divided their world into friends and enemies, drew arbitrary boundaries around themselves and responded with quick emotion to threats from outside the gene pool or the immediate ecology. Aggression was expressed, became formalized and entered easily into our repertoire of behavior because the ones who were best at it became the most successful. Nothing could have stopped that tide from coming in. Anyone who tried, died; while the warmakers and warmongers prospered. They still do and now we find ourselves on the brink of Hobbesian catastrophe in a lawless chaos which does put "all against all."[187]

It has happened before, on a small scale. A thousand years ago, a Polynesian people on Easter Island, perhaps 7,000 of them, raised hundreds of giant statues on temple platforms looking directly out to sea in a powerful advertisement of their culture and of enviable peace and order. Then something went wrong, the climate changed perhaps, and power was assumed by a warrior class—the *tangata rima toto,* the "men with bloodied hands." They toppled the statues, dug defensive ditches, fell out with one another and fought so incessantly that by 1722 there were just 111 people left on Easter Island, with only the sketchiest memory of their golden age.[120]

They have little to teach us except how to die, but there are others still relatively untouched from whom we can learn much of the wisdom of ritual and restraint. Something that might still help us pull back from the brink, before it is too late. The sad truth, however, is that the odds are not in our favor. As Abba Eban remarked ruefully during the 1967 Arab-Israeli war: "We are but men, and use reason only as a last resort."[426]

There is some justification for that bleak assessment. Aggression is instinctive to us in the sense that, when faced with unrelated members of our own species, we are more likely to be nasty to them than nice. The genes encourage cooperation only among close kin. And we are most likely to harm others at the time of our lives when we are also most concerned with reproduction. That is when the genes' concern about their own success is greatest and most likely to be reflected in competition for suitable mates. The fact that such competition usually occurs between males is well reflected in human statistics. It is reinforced also by the fact that only men have a collection of genes grouped into what has been called the Y chromosome, and that having two such chromosomes—a mutation that occurs in perhaps one man in every thousand—does seem to make such an individual more liable to express themselves aggressively.[258] It is true that males in our species, as in many others, are far more likely to injure or kill than females. And the way in which they do so, their patterns and rituals of aggressive behavior, tend to have a local flavor. In all these depressing respects, aggression can be said to be biological, inherited rather than learned. But that is by no means the whole story.[30]

We also know that aggression occurs in a variety of forms, ranging from simple defense to open war, and that in all these its expression is, at least in part, dependent on external conditions. We are touched not only by our collective biology, but by our individual history. If the last time you met someone, you fought, then the chances are that the two of you may well do so again. A lot of violent behavior is self-fulfilling in this way and spills easily over onto others in our societies. Revenge is sweet even for Japanese macaques, who are known to redirect aggression onto the kin of the monkey who has just attacked them.[18] The larger the brain, the more complex the society, the more likely such strategies are to multiply and to muddy the social water. But the very fact that these are cultural patterns suggests that they need not be inevitable. Large brains are capable also of finding ways to exercise and encourage restraint. Only ants are condemned to automatic, unthinking combat between rival colonies.[205]

Human wars may be unreasonable, but they do not take place because of a failure to bring reason to bear. Far more often they are the result of a deliberate negation of reason, of a rational decision by someone to glorify a cause and to provoke others into defending or attacking it. This has little to do with biology, and everything to do with economic, political and social factors—the "mental furniture" of the special place at a particular time.[178] Aggression is part of our furniture, but we have a choice as to how it should be arranged. War is no more intrinsic to human society than slavery. Both can and should be seen as aberrations of our nature.[431] Both also, however, raise questions about other aspects of our behavior that appear to be maladaptive enough to be regarded as "wrong" or "bad" in ecological and evolutionary terms, perhaps even as true evils.

Most emotive and disruptive of these is the question of rape, of copulation by force, by a male, against the will of a female. Such behavior is common, but is not peculiar to our species. The male scorpion-fly we met earlier as a bringer of nuptial gifts and occasional deceiver will also sometimes take a female fly by force. If a male cannot find a suitable dead insect, or is not quick enough to steal one from a spider's web, or hasn't himself eaten recently enough to secrete a salivary imitation, he may simply lie in wait for an unwary female. As she passes, he lashes out with his flexible abdomen and tries to grasp her leg or wing. Most of the time, she struggles free and flies away. But if he can secure a hold, he transfers this and clamps himself to her right forewing, while repositioning her so the clasper on his abdomen becomes attached to her genitals. And he holds her, still struggling, in this position throughout copulation.[375]

This is rape, by any definition. But it is not pathological or aberrant. It is a part of the normal behavioral repertoire of scorpion-flies, a strategy available to any individual male at any time, though only usually as a last resort. It is most often employed by small and less successful males and is vigorously resisted by female flies, who always lay fewer eggs after forced copulation. The clear preference of the females is for larger males who come with proof of their fitness and whose advances are met either with behavior that gives every appearance of coy acceptance, or with open submission that permits a full and leisurely transfer of all the male's supply of sperm. The combination of male fitness and female choice is clearly the best strategy for scorpion-flies, the one which offers the greatest benefit to their genes, and which produces offspring that have the best chance of being equally successful and selective. Anything less than this is second or third best, a disaster as far as the female is concerned; but for a male of low fitness, it may be better than nothing at all. A viable alternative for a loser.

Evolution tends to exploit every avenue available to it, giving up only on those that produce no advantage at all. In Darwinian terms, there is a reason for rape, a slim arithmetical edge that gives it a place in natural history. This does not, of course, justify such behavior, but it does help to arrive at a model which puts the action into biological perspective.

Presumptions, myths and misunderstandings about rape abound.[46] It is thought by some to be a kind of social pathology, a disease of industrial civilization; but the evidence suggests that rape is crosscultural, an unfortunate human universal. It is believed to be an act whose sole aim is to dominate women; but there is no evidence to show that rape is directed primarily at rich, powerful, older women. On the contrary, its targets are usually poor, helpless, younger women. And if it is a propensity of all men, they ought to be equally represented in the crime; but the vast majority of rapists are quite clearly poor, young men.[377]

Biologist Randy Thornhill believes that the scorpion-fly example, while not directly comparable to human behavior, is useful in pointing to a similar evolutionary strategy in our species. He suggests that, whatever else may be involved in acts of rape—and there certainly are psychological, political, behavioral and racial factors to be considered—there is a reproductive link. It is no coincidence that most victims are young, nubile women at the age of maximum fertility; or that rape should be the only crime against women that is directed specifically at this age group. And he predicts, on this basis, that the men most likely to rape will be "losers," those who have failed to climb the social or economic ladder and have the greatest difficulty in attracting desirable mates in the normal ways.[376]

The prediction seems to be true of most individual human rapists who, like their smaller, weaker scorpion-fly counterparts, are of low reproductive fitness and self-esteem. In their circumstances, rape—despite all its risks and costs—might still represent the only viable genetic alternative for an individual who cannot compete successfully. In any less complex species than ours, this would be full and sufficient reason for a tendency to have evolved and survived. But in humans, where rape may not be so much an aggressive manifestation of sexuality as a sexual manifestation of aggression, it is not so easy to define.

Rape is unquestionably a violent act, one that degrades and humiliates its victims, but it would be a mistake to ignore the fact that it also makes some of them pregnant. Where the genes are concerned, it is always a mistake to ignore any possibility that involves possible reproduction. That is their business and they are very good at it. There is a valid argument to be made for looking at rape as a maladaptive residue of an old pattern that did once have survival value in biological terms. Now it does little but make a battlefield of women's bodies, condemning them to be not just the objects of repressed desire, but the targets of rage against all women who are seen to "belong" to other better-endowed men. And nowhere is this more obvious than in war.[47]

War is a game with its own rules, which seem to supersede those of civilian society. Many of the rules are unwritten and by far the most common of these is one that takes for granted that rewards for a successful campaign will include women. The Yanomamö make no secret about it, they fight to get more women. "To the victors belong the spoils of the enemy"—and the spoils most soldiers still look forward to are feminine.[259] No nation or army is immune, even if some seem a little coy about it. The best remembered and most often reproduced image of the Second World War was not a battle scene or a flag going up somewhere, but *Life* magazine's giddy shot of a sailor kissing a girl, who had probably never seen him in her life before, during a victory celebration.

War is an open space, a time out of mind, when anything goes and atrocities often do go unpunished and even unrecorded. There are conspiracies of silence surrounding many such events that at least imply guilt and a sense of shame, but at the time, morality is put into abeyance and men in uniform, any and all uniforms, routinely debase women as a way of enhancing their own masculinity in each other's eyes. Group rapes, gang rapes and platoon rapes all take on the structured form of exercises with a personal weapon in which the victims become faceless, almost incidental to a maneuver designed to reinforce the military construct of men doing manly things together.

The results are horrendous. In 1937, when Japanese forces captured the Chinese city of Nanking, more than 20,000 women were sexually tortured and murdered in a single month. Truly the "Rape of Nanking."[135] In 1943 Moroccan mercenaries fighting with the French were given explicit license to plunder and rape their way through tens of thousands of Italian women.[404] And in the spring of 1945 in liberated Berlin Allied troops took leave of the war and of their senses, raping hundreds of thousands of German women, included among whom were victims of Nazi concentration camps.[342]

Official response to all these outrages has been ambivalent. They played little part in war crimes investigations and nothing was ever said or done to deny that blind eyes were turned from on high by the officers of each army at the time. Such things seem to be a part of the philosophy of war, part of a deliberate program to destroy the enemy's culture by humiliating him and impregnating her. Studies show that many of the husbands of the victims of the Berlin debauch ended their marriages as a direct result of the rapes, which laid waste to their land and their most precious "possessions."

And still it goes on. In 1971 more than 200,000 Bengali women were raped in Bangladesh as part of a conscious military strategy by Pakistan. In former Yugoslavia, the Serbs set the pattern going again in 1992, practicing systematic rape in camps set up for that explicit purpose, violating children in front of their parents and on camera with cruelty so callous it makes just killing seem kind by comparison. The Croats responded in kind in 1993 and, as far as anyone can tell, such atrocities continue behind political smokescreens set up to direct attention from the horrible truth—which is that old scores are being settled by the thousand, by all the ethnic groups involved, on everybody else's women as a historic byproduct of war.[367] A direct modern extension of the langurlike instructions given by Moses to his troops after the battle against the Midianites: "Now therefore kill every male among them, and kill every woman that hath known man by lying with him. But all the women, that have not known a man by lying with him, keep alive for yourselves."[193]

There are genetic influences at work here, but also a "pressure cooker" theory of male nature, which presupposes an uncontrollable sexual drive that runs wild in time of war, when it is no longer restrained by the civilizing influence of women. Perhaps, but this sounds too much like a convenient excuse for behaving very badly under stress. There are no acceptable excuses for rape or torture at any time, at least not by human beings who may be driven by the same primal urges that move langurs and lions to infanticide, but have enough independence now from genetic pressure for those involved to know better. Consciously, at least.

The difficulty we find in behaving well, in doing the "right thing," the one that represents the Goldilocks solution, is compounded by our alienation from nature in war and social upheaval. The "military horizon" seems to me to coincide with a critical environmental quantity, a degree of natural diversity and interconnection that we cannot properly live without. Take that away, go beyond that horizon, and we are cast adrift, rudderless, doomed to be bad, to practice evil simply because we can no longer remember what is right and what is wrong.

That way madness lies. There may be some way of working around the dilemma, of growing old and growing up without sacrificing some of the comforts of growing wealthy. But before we look for such recipes, and if we are ever to be fair to each other and to our nature, we need to look harder at a few more disagreeable facts about ourselves.

FIVE

THE WAGES OF SIN: THE PSYCHOLOGY OF EVIL

It is difficult for an untrained, unarmed human being to kill another. We are not designed for it. Our claws and teeth just aren't up to the job. Compared to most predators, we are woefully inadequate; and yet we have managed to become the world's most dangerous animal, thanks to technological accessories that carry us well beyond organic evolution. Culture has swamped and overtaken basic biology, giving us an extraordinary capacity for mutual destruction, without the inhibitions that accompany most carnivores' possession of their formidable natural weaponry.[368]

Real violence is rare in nature, a last resort. Most species settle for something less. Sometimes a threat will do, even one made by proxy. Grizzly bears put their territorial claw marks as high up the trunk of a tree as they can reach, standing on tiptoe, giving intruding males pause for thought. And even when rivals meet face to face, there are rules which govern such encounters and these are remarkably unambiguous. Each participant knows exactly what constitutes aggression and, just as important, what constitutes submission—and both know where to draw the line and how to walk it with precision.

Our species specializes in being unspecialized. All we really have going for us is a restless exploratory nature and, given our lack of personal arms, an astonishing amount of self-assurance. We are the apes with attitude. Pushy primates with just enough of the right kind of aggression, but not quite enough of the inhibitors necessary to keep us out of trouble. We seem to take pleasure in taking risks, finding fun in testing the limits defined by the line between aggression and submission—even if it sometimes ends in tears.

"A little bit of violence never hurt anyone" according to a recent graffito on the streets of London. And strangely, that may well be true. British psychologist Peter Marsh makes an important distinction between real violence and what he describes as the illusion of violence. "Because we don't understand aggression," he suggests, "and because we fail to realize that it might have a variety of consequences—some of which can be socially useful—we find it increasingly difficult to manage the process." We miss the point, we lay blame in the wrong quarter and end up with something far more sinister. The constructive social violence he has in mind is *aggro*—a uniquely British term for a widespread technique of expressing aggression in a relativeless harmless way.[262]

"Aggro" comes from aggravation and has its origin in the sort of boundary and dominance disputes that leave rival male chimpanzees with all their hair erect, hooting, screaming and glaring at one another, arms waving over their heads, and feet stamping on the ground. All that description lacks to make it an accurate one of human behavior in the same circumstances is the addition of a rich repertoire of vocal insults and some appropriate flags or banners.

At the Circus Maximus in ancient Rome, 200,000 spectators (twice the number possible at Wembley or the Houston Astrodome), proudly sporting rival colors of blue or green, cheered their favorite charioteers on to success or failure, occasionally coming to blows with one another. These were truly classical hooligans, fighting rival ends, setting up clubs and factions, establishing territories and feuding right through into Byzantine times.

Reaction to them was as predictable then as conservative outrage today is toward the rowdy behavior of sports supporters. But what the Roman authorities understood a little better, perhaps, was that such aggro was no threat to the *status quo.* On the contrary, rumbles at the race track, or a bit of "bovver" at the circus served to let off steam and were so carefully structured that comparatively little harm came to anyone. Behind all the bluster, the costumes and the customs, there are unwritten and largely unconscious rules that govern encounters, even in modern aggro. Supporters of football teams such as Liverpool or Millwall, for instance, despite some sensational lapses, are not mobs but true groups, with shared values and a surprising degree of internal constraint, taking part in a complex social process which includes mechanisms for handling aggression reasonably safely. Sometimes the rivalry on the field does spill over into more dangerous antagonism during or after the game, but on the whole it is a technique that works remarkably well, ritualizing and channelling inevitable youthful aggression away from riot and toward the common good.

In the United States a persistent frontier feeling seems to play a more prominent part in such aggression, complete with the sort of weapons appropriate to the period. Starting in the 1920s, street gangs such as the infamous Vice Lords of Chicago, the Spanish Kings of the Bronx, and the Crips in Los Angeles, carved up most inner-city areas into a matrix of zones, each of which constituted defensible "turf."[226] But unlike animal territories, which are specifically designed to arouse aggression, these home areas appear instead to provide some social control of it. Beyond the threats and graffiti which decorate the range, behind the "colors" that the members wear, is a kind of self-enforced discipline which allows street warriors to act out their inherent aggression in a relatively safe and orderly way. The turf is their parade ground, their place to strut, an arena in which they can "look good" in front of each other and in front of the majority of ordinary people who live in such neighborhoods.

The gangs there were never fighting to protect such little empires; they establish these stamping grounds only because they already want to fight—and what the territories do is to provide the structure for shaping such combat. For ritualizing it in ways that are a lot less lethal than when aggression has to work itself out without the benefit of symbolic lines drawn in the urban sand. The gangs certainly hurt and kill each other, but it can be argued that they are not so much involved in violence for its own sake as in the management of violence by encouraging more ceremonial forms of conflict resolution.[434]

Gang power is beginning to wane as mere anarchy consumes even the best organized among them. But they are still the masters, as Peter Marsh puts it, of ritual restraint: "Of the swinging fist which *just fails* to connect with anything."[262]

Traditional violence has always been carefully regulated. On Tory Island, a rocky, windswept community off the west coast of Ireland, there used to be "fights" several times a week outside the local church hall. Protagonists began these ritual encounters with long and highly creative exchanges of insults, following which one man would allow the struggle to escalate just enough to permit well-simulated rage in which he threatened to kill the other, unless he was restrained. He demanded that his friends and relatives "hold him back," and that is precisely what they did, effectively preventing him, despite all his histrionic struggles, from taking his jacket off. That would be too much, a potent signal in the form of a recognized intention movement that always precedes true fighting. Making it commits both parties to actual physical combat that in a small community could be socially and genetically disastrous. But the well-rehearsed and closely related crowd seldom allowed things to go that far.[130]

In urban areas today, however, such restraints have largely disappeared. We have sacrificed appropriate social expressions of aggression such as aggro, and put nothing in their place. We are not born killers of our own kind, but without tribal custom and concern, without even the protection of tribal substitutes such as the once powerful street gangs, our jackets are well and truly off, and we have become involved instead in a war of all against all—a struggle in which no holds are barred, and no one is immune.

Violence appears to have become epidemic. Random killings in schools, hospitals and restaurants; isolated car-jackings that can take place at any suburban street corner; senseless drive-by shootings with no apparent motives—all these feed a new and widespread level of personal fear. Morton Hunt, in a vivid story called *The Mugging,* suggests that: "What alarms us and most gravely damages our faith in our society is the ever present threat of some sudden, unpredictable, savage assault upon our own body by a stranger—a faceless, nameless, fleet-footed figure who leaps from the shadows, strikes at us with his fists, an iron pipe, or a switch-blade knife, and then vanishes into an alley with our wallet or purse, leaving us broken and bleeding on the sidewalk."[204]

Mugging is now the most common of all crimes involving physical force, despite the fact that it yields an average reward scarcely high enough to meet the cost of a cheap meal in a fast-food outlet. The theft seems almost incidental, providing little more than a convenient focus for violence that now promises to bring the new anarchy so close to home that most people in the United States, where one in every 300 have been mugged, have turned their homes into fortresses where they undergo virtual self-imprisonment. The fright is real and realistic, even if exaggerated somewhat by media attention which feeds the perception of mounting chaos, and makes us all as obsessed by violence, and by thoughts of violence, as our Victorian great-grandparents were about sex. Like them, we are "outraged and appalled," but we are also fascinated and vicariously aroused, manufacturing and enjoying new and more devastating images of violence with each increase in the budget of feature films, and importing news of violence from abroad when there isn't enough of it on the domestic television scene. We are building a "culture of violence" in which such images have become difficult to ignore. Escapism is a lost art. No one in their right mind goes to the movies, watches television or listens to popular music these days to get *away* from violence. There is no avoiding harsh reality—or the current violent perception of reality—short of entering a closed order. And even then the monastery or convent is likely to be surrounded by a spiked fence topped with razor wire and protected by Doberman pinschers.

Paranoia about violence is rampant, but as sociologist James Wilson points out: "If you map the fear of crime and map the actual range of crime, you note that they don't overlap."[428] This may be a direct result of suburban concern producing the sort of neighborhood security that really prevents crime. Most urban communities cannot afford to do that, but have found another social ploy that meets their needs in a surprising way. In the inner cities, street culture has evolved its own set of rules to live by: "The Code of the Streets."[8]

These rules are to do with respect and how to get it. They prescribe proper comportment, the "right" way to behave and the approved way of responding to the behavior of others. Knowledge of the code is essential to survival, and everybody is held responsible for such familiarity. Failure to do the right thing can be lethal, so the code is essentially defensive, even chivalrous, in its nature. It regulates the use of violence and allows those likely to be aggressive to precipitate violence only in an approved way. Part of the code deals with appearance, taste is tightly regulated, but mostly it has to do with bearing—with facial expression, language and gesture—all of which have to be geared not to producing, but to deterring aggression.

Despite everything you may have heard about the attitude of inner-city kids who "come up hard" and are obsessed about being "dissed," they seem to have engineered a solution to life on the mean streets that works. And it is largely individual. This is not a code enforced by gang pressure. These are a new set of rules to live by. Rules that make it possible for anyone to live through anarchy which society as a whole seems powerless to control.

The code, whether it is practiced in Calcutta or Lagos, Palermo or Harlem, produces the same heightened sense of urgency, something appropriate for those already living on the edge. It takes nerve to practice properly, as well as constant vigilance and an acute awareness of nuance in the bearing and behavior of others. Campaigning for respect is a political process as finely tuned as a well-played game of Tit For Tat. Everything depends upon making the right moves at the proper time, on defending yourself when challenged, and on striking back when that becomes appropriate.

It all sounds very difficult, but in fact this is a way of life for which we have been in training for ninety-nine percent of human history. Keeping up with all the variables, "keeping yourself straight," finding the proper balance between aggression and restraint, is something we know how to do from long practice, even if it still makes us as jumpy as monkeys working out the finer points of constantly shifting relationships in a new troop's social order. Men and monkeys under such circumstances become very thin-skinned, craving respect to such a degree that they may even risk their lives to attain it. But despite the obvious danger inherent in such hair-trigger brinkmanship, and the fact that lethal mistakes can be and sometimes are made, the system actually works very well, cutting down on random aggression both on the Rock of Gibraltar and in the streets of a Chicago ghetto.

There is, in the evolution and existence of such patterns, which appear and reappear precisely where and when they can be most effective, a cause for optimism, a reason to hope that we can find enough ammunition in the arsenal of natural history to deal with almost any eventuality. As a species, we are nothing if not inventive, but there remain some areas of life in which we seem to have painted ourselves into corners from which it may be impossible for everyone to get out alive.

I have already suggested that aggression needs to be seen as a given, as a natural, useful and inevitable part of our genetic inheritance. It is one of the things that has made us human. It comes of course in many forms, most of which are "instrumental"—they serve some goal and are purposefully used to achieve that end. Even the fierce Yanomamö stop short of killing enemies who have already been immobilized and therefore represent no immediate threat. But there remains a residue of a different, more impulsive, kind of aggression.

Impulsive anger seems to play little part in war, where most men fight and kill, not in anger, but out of fear of letting down their fellows. A survey of American combat troops during the Second World War found that only twenty-five percent would use their weapons directly against a visible enemy, even when hard pressed. The great majority of casualties in modern war come from weapons used at a distance, where the target becomes impersonal, even invisible, and has little more than symbolic value.[77]

This distancing began perhaps 5,000 years ago with the Sumerian invention of the composite or laminated bow, capable of shooting an arrow over 1,000 feet, and was followed perhaps 2,000 years ago by the Chinese perfection of the armor-piercing crossbow. Both weapons are lethal and produced quantum leaps in casualty figures, but were strongly resisted by aristocratic warriors who considered them unsporting and ungentlemanly, largely because they were egalitarian.[264] Such weapons could be used by almost anyone, even a woman, making nonsense of the hard-won skill with edged weapons that made medieval knights-at-arms and Japanese samurai, all drawn from the male nobility, so important to the old tradition of face-to-face combat. The great French knight Bayard—*chevalier sans peur et sans reproche*—habitually had prisoners executed if they were known to be crossbowmen, on the grounds that this weapon was a cowardly one and their behavior treacherous.[225]

The advent of the bow, and later of the firearm, effectively brought the age of chivalry, in which battles were fought under strict rules, to an end. The code of the warrior, of never being the first to show aggression, seems to have survived only in the cinema western. But guns are still very much with us, and they present a real problem, making impulsive anger a major factor in our affairs by lowering the threshold at which such incidental aggression can now become lethal. There are no traditional inhibitions, no biological or social restraints, no natural defenses, against the gun; which turns every child into a superpredator with the power to kill at will.

All it takes now is the twitch of a finger. Sometimes even that isn't necessary. Hundreds of people are killed each year by firearms accidentally discharged by domestic pets. The principal culprit in this carnage is the handgun—twenty of which are manufactured every second in the United States alone, and put into 71 million hands and homes. Each of these instruments is designed, cast, calibrated and sold with just one purpose, to end human life, which they do very well. There were 13,220 handgun murders in the United States in 1992, just 262 of which were ruled as justifiable homicides in cases of self-defense. Handguns were involved in sixty percent of all homicides and seventy-three percent of the murders of children under the age of fourteen. And each year the statistics get more terrifying and less logical. The technology of the guns and their ammunition is becoming ever more refined, making them more lethal in every catalogue, where their efficiency in this respect is used as a selling point. It works. A million violent crimes a year are committed with handguns and in just three years, between 1990 and 1992, more Americans were killed by them than died during the entire seven years of the Vietnam War.[220]

The year 1992 was not unusual. It is simply the most recent set of figures I could find for the United States. But America is unusual in global terms in allowing handguns to be bought and owned by anyone except the manifestly lunatic and the criminally convicted. Other nations have stricter standards for gun ownership and these are reflected very directly in their figures for handgun murders in 1992. Against 13,220 in the United States, set 128 in Canada, 60 in Japan, 36 in Sweden, 33 in Britain, and just 13 in Australia. That means something. But to put this slaughter in perspective, we need to take a look at homicide as a human habit.

Murder fascinates us. We spend a disproportionate amount of time, money and energy in reporting, analyzing, solving, prosecuting and trying murder cases; and trying, almost as an afterthought, to prevent more of us from killing one another. "True crime" and first-person accounts of real-life homicide cases have become a growth sector of modern publishing, threatening even to outsell more traditional murder mystery fiction. Departments of criminology and psychopathology exist now in most major universities and financial support for further research appears as a regular, knee-jerk item in national programs designed to combat crime. But we still have only the most rudimentary understanding of who kills whom and why.

Killing members of one's own species is the ultimate technique in conflict resolution, an act of desperation for most animals, who shed each other's blood in combat with reluctance and against powerful inhibition. We appear to be less reluctant to do so, and not because of rare and psychopathic disorders. A few isolated individuals do kill people and take pleasure in eating parts of their bodies, but most murderers appear to be ordinary people, exercising what looks like a "normal" part of our behavioral repertoire. And if this is so, we need to know how things came to be that way.

A significant proportion of murders take place in the home, so often involving only members of the immediate family that one study concluded: "Violence is so common in the family that it is at least as typical of family relations as is love."[139] Part of the reason for this pattern, which goes directly against the genetic instruction to be nice to insiders, may be that family members are most often targeted simply because they are there. As sociologist William Goode points out: "We are violent toward our intimates—friends, lovers, spouses—because few others can anger us so much. As they are the main source of our pleasure, they are equally a main source of frustration and hurt."[154] And the proximity and exclusivity in which we find ourselves thrown together now is artificial in the sense that "one-family homes" are a very modern invention. But the most interesting part of this line of research has been the discovery that some members of the very limited pool of people available in such a home are more at risk than others.[280]

Some of the best, most comprehensive and most sensible studies on homicide in recent years have been made at McMaster University in Canada by Martin Daly and Margo Wilson, psychologists who bring welcome evolutionary insights to their work. And they have found, among many other things for which I am in their debt, that unrelated family members—such as spouses, who are not blood kin—are eleven times more likely to be killed.[89] And the main difference between the homes where this happens today and those of the larger extended family more common in historic times seems to be that killings in those days were more likely to be collaborative.

The rolls of court proceedings in thirteenth-century England show that domestic murders were often done by two or more adults, and that these conspirators were invariably blood kin acting against another adult who was not.[146] That makes genetic sense and puts such domestic homicides firmly into a long human tradition of family quarrels and blood feuds. Our urge to kill is at least partly situational. It happens so often in the home for the same reason that most motor accidents occur within twenty-five miles of the home—this is where we and our cars most often *are.* And once the urge to kill exists, genetic factors come into play in the choice of a victim. We are far more prone, six times as likely by most estimates, to be terminally nasty to outsiders. And where the victim does turn out to be an insider, there are very often factors involved that make biological, if not moral, sense.[94]

Humans are among the most intensively parental of all species, joining whales, elephants and apes in maintaining bonds that can last for decades. In prehistoric times, a woman was lucky if she could rear two or three children successfully to maturity, and able to do this only if she made hard decisions along the way about her priorities. These, as we have seen, may have included neglecting or even killing a newborn child to ensure the well-being of another older infant. Or, where paternity may have been in doubt, sacrificing a child to satisfy her mate's doubts and to ensure his future cooperation and investment in rearing children he *did* regard as his own.

Other factors sometimes leading to infanticide include deformity; being a twin; being born at the wrong time of the year in a nomadic society committed to making a major move; being a girl when a boy child was valued more highly; being a boy unlucky enough to be born at the same time as the chief's son and heir; or simply having an unmarried mother. Daly and Wilson list 112 such "infanticidal circumstances," many of which persist and would be considered unsurprising even today in communities where decisions continue to be made in the interest of inclusive fitness. In such cases the custom is legitimized by that society, but it is one made, often with reluctance, directly by the parent or parents themselves, because it is seen to be in the best long-term interest of the family and its inheritance.[90]

Much of the early literature about such child-killing is ethnocentric and inappropriately moralistic. It is actually very hard to condemn practices which have such strong genetic backing and make ecological sense. As with headhunting, I suggest that we look again at the context in each case without rushing to judgment and dismissing all infanticides out of hand as evil.

On the night of October 25, 1994, Susan Smith put her Mazda car into gear, slammed the door and let it roll down the boat ramp into a dark country lake in South Carolina. Strapped into the back seat were her two baby sons, the ones nicknamed "Precious" and "Sugarfoot." An hour later, she accused a black man in a woolen cap of stealing the car and kidnapping her children at gunpoint, starting a manhunt that had thousands searching in response to her tearful appeals on national television. But when nothing was found, the investigation turned back on the immediate family and Susan made a stunning confession. She told the police where the car and the boys' bodies could be found.[141]

The deaths, and the deceit involved, produced shock and outrage in a nation accustomed to explaining the danger of strangers to their children, but faced now with the far scarier statistic that, in the United States alone, 1,300 children are killed each year by their parents or close relatives. Half of these victims are under the age of one, and most of these are murdered by their mothers.[256]

Infanticide is nothing new. It is not an artifact of modern times, but has always been one of the most common forms of homicide, putting babies at greater risk than any other part of the population. Mothers who kill their children are, in fact, almost ordinary—nothing like the image conjured up by Lady Macbeth proclaiming: "I would, while it was smiling in my face, have plucked my nipple from his boneless gums, and dashed the brains out."[345]

Susan Smith's case is typical. Pregnant at nineteen and married in haste. Estranged at twenty-three after she and her husband had other affairs. Unable, despite child support, to meet her debts, spurned by a well-to-do lover who was unwilling to accept her children. With a history of unsuccessful suicide attempts, she was a candidate more in need of compassion and mental health treatment than the jurisdiction of the criminal law. But that is where she now resides, serving a life sentence along with others such as Dora Buenrostro who stabbed her three children to death in California; Waneta Hoyt of New York, who has confessed to suffocating all of her five babies; Renee Aulton, who started a fire that destroyed her home along with two infant daughters; and Elizabeth Downs in Oregon, convicted of shooting her three sleeping children. The list is long and growing, along with a suspicion that confusion, delusion and radical shifts of mood and perception may be common and surprisingly persistent side-effects of postpartum depression.[250]

These examples are ones that cannot be described as contributing to social stability and ecological equilibrium, but neither can they, nor should they, be put beyond explanation as outbreaks of unimaginable evil. We have to be careful not to confuse the interests of parents and offspring, which often conflict where optimal fitness is concerned. Children nearly always want more than their parents can provide, and nice judgment is required to reconcile such disparity. In many situations, unconscious calculations are clearly being made, with every evidence of an evolutionary perspective coming into play.

Most abortions and most cases of infanticide involve teenage girls. But as mothers grow older and their chances of reproductive success grow slimmer, they show a growing and more marked reluctance to terminate pregnancies or to put infants at risk in any way. That too makes genetic sense. And as the balance shifts, and children find themselves in the position of having to take their own hard decisions, aging parents become increasingly more likely to be killed by their offspring. The older and needier they are, the more easily they can become a liability, reducing their children's inclusive fitness. Not all offspring resort to parricide, of course, some can afford geriatric care, but even those who choose to accept responsibility for their parents find that the cost can include conscious or unconscious resentment produced by an awareness of their own family's reduced circumstances.[92]

In these unbalanced family affairs, one further trend is becoming clear, and comes as no surprise. Substitute parents, those without genetic investment in the children, have a poor reputation. Stepmothers are traditionally "wicked" and uncaring; stepfathers "cruel" and lustful. Children tend to be deeply suspicious of both in many cultures, perhaps with very good reason. In the United States, forty-three percent of all children so badly abused that they died were living at the time with stepparents.[91] And in England, an astonishing fifty-two percent of all babies killed in the home were beaten to death by stepfathers.[341] Data from Canada, which also take the child's age into account, show that lethal "baby-battering" is sixty-five times more likely to take place in a home that includes a new stepfather, who may have a langurlike antipathy to another man's genes.[50]

It is obvious that the parental role is hard to play. Nature makes it easier by encouraging biological parents to care profoundly and selflessly for their offspring, no matter what it costs. That is a hard act for any substitute parent to follow. There is bound to be a disinclination to make such a substantial investment in someone else's child, particularly in one predisposed to regard you with suspicion. But there can be no excuse, regardless of genetic influence, for a man to beat a child to death in temper or frustration. That is unforgivable. An unquestionably evil act—not by its nature, but by the moral grounds on which humans, unlike all other species, need to be judged. This is one of the burdens of self-awareness, something I want to discuss in more detail later.

For the moment, there are other homicidal tendencies to explore, including ones arising from Sigmund Freud's contention that every boy has the urge to kill his father and sleep with his mother. As Daly and Wilson point out, there is no evidence in any data on homicide to support the Oedipal theory, no significant trend anywhere to show that sons are killing their fathers any more often than young men, who tend to be the most violent group, are killing older men in general, who tend to be the ones who own the things worth stealing.[93]

There is evidence of resentment toward fathers by both sons and daughters, but this seems to be simply because children are not anxious to be weaned, and are inclined to be hostile to new pregnancies. "From a child's point of view, daddy has his own uses for mommy, and they are not necessarily harmonious to the child's. Parental sexuality threatens to produce a younger sibling, and it is not implausible that even young children should evolve specific adaptive strategies to delay that event by diminishing mother's sexual interest and by thwarting father's access to her."[94]

This is not exactly what Freud, who read but apparently failed to understand Darwin, had in mind. This is just normal conflict between the generations which has nothing to do with sex, and reaches its peak, prompting a brief flurry of patricide, when parents who hold family resources appear reluctant to pass these on to maturing children. Statistics show that patricide and matricide increase rapidly as both parents get older and become more of a burden.[92] These are global trends and evidence of a distinct and growing asymmetry between the interests and the needs of the generations, only one of which continues to have a genetic investment in the survival and wellbeing of the other. Hence the pressures everywhere on children to get married and to breed as soon as humanly possible, and a widespread tendency by parents, when disputes do occur, to skip a generation and pass material inheritances directly on to their grandchildren.

Outside the family, homicidal data become much more simple and universal. The evidence everywhere is that it is men who kill, and that they most often kill other unrelated men, usually in struggles for status and respect. There is no known human society in which this is not true. Such lethal violence by women is, by comparison, extremely rare. One in every four murder victims in the United States is a woman, and nine out ten murdered women are killed by a man. Four out of five are murdered at home, three of those by husbands and lovers. Hardly any are killed by strangers. Women who do kill, on the other hand, nearly always kill their husbands, lovers or children. Those with whom they have been most intimate. And there is no evidence, despite antifeminist sentiments, that murder committed by women is any more common now than it was before women's suffrage. The numbers of women in prison may have risen rapidly in recent years, but they tend to be there on charges of larceny and theft, not for violence or homicide, and are being put there by an increased willingness of police to press such charges, regardless of sex.[213]

The fact remains that there are clear sex differences in human behavior which appear to have nothing to do with environment and cultural expectations, and everything to do with basic biology. Daly and Wilson have shown that in every society for which information exists, competition between men is far more violent, on average twenty-four times more violent, than competition between women.[94] And, as usual, there is a biological ground for this disparity. Men have a lot to gain and lose in such competition. The winners get women, the losers don't. And the biggest winners get the most women and an entry in the *Guinness Book of Records.* Moulay Ismail, Emperor of Morocco at the beginning of the eighteenth century, earned his sobriquet of "The Bloodthirsty" by killing an estimated 30,000 people with his own hands, even lopping off the heads of slaves to entertain his guests at dinner; and he had 500 wives, who gave him more than 1,000 offspring, of which he formally acknowledged 888.[254] There is no record of the number of Moroccan losers who, as a result, failed to find wives or to produce any children at all.

In such competition, there is a very small chance of winning big, but there is a very large chance of losing altogether. Such odds favor the taking of necessary risks. Men do, and their tendency to do so depends directly on their status. The lower your status, the greater your gamble will need to be to enter the reproductive stakes, and the more likely you will be to look at your prospects in the short term. Studies of young black men in inner-city environments in the United States show quite clearly that the ones with the most "juice"—the most admired by their peers as "bad dudes," who also prove to be the most attractive to young women—all live very close to the edge. They don't expect to live long and lead an existential life that seems to take on meaning only when they are faced with the possibility of sudden death. And this apparent lack of fear of dying is, not surprisingly, accompanied by a relative lack of compunction about killing.[8]

The result is predictable. In the United States and Canada, over ninety-five percent of all homicide victims and offenders are male, mostly in their twenties; over forty percent of both are unemployed; and over seventy percent are unmarried.[430] The "motive" on police blotters and charge sheets is usually recorded as "Altercation of trivial origin," but that merely obscures the fact that such disputes are invariably about respect of the kind likely to lead to enhanced status and greater reproductive success. The genes and their needs play a prominent part in most such social encounters that end in homicide, but begin with men looking for resources which exceed their immediate needs for food and shelter, while meeting their equally compelling need for things like clothes, cars and other accessories likely to enhance their sexual attraction.[383]

The problem doesn't, of course, end with sex. To ensure successful reproduction as a result of sex, most human societies formalize a breeding relationship with some kind of alliance or marriage which usually means exclusive sexual access. And this, more often than not, leads to men all over the world thinking about women in proprietary terms and exercising extraordinary double standards.[429] The very word *adultery* betrays the bias. It derives from the Latin *adulterare,* which means "to make impure, spurious, or inferior by adding extraneous or improper ingredients." In other words, it remains—as it has through all of animal behavior—a question which touches on male doubts about paternity. With these doubts comes righteous anger and sexual jealousy sufficient to account for the vast majority of wife beatings, and most of the murders of women in marriage. And the statistics show an additional blip in the case of marriages where the husband is significantly older than the wife and has good reason to be fearful of younger and more potent male rivals.

This confusion of women with property spills over into revenge killings by men who cannot easily let women go. Every police file includes cases of estranged wives or girlfriends who have been hunted down and murdered, but there are very few known examples of vengeful murders by jilted wives or lovers. The few that do come to light make news and controversial feature films such as *Fatal Attraction.* There are elements of spite as well as jealousy in such homicides, because a third of all men in a Canadian study of those who killed estranged wives also killed themselves. And several, apparently obsessed with possible infidelity which brought their paternity into doubt, killed their entire families as well.[159]

Familicide is perhaps the most horrifying aberration of genetic pressure, leading someone to carry out a course of action which is spectacularly futile and devastating to his or her own interests, just for the sake of inflicting damage on someone who appears to have put an investment at risk. Such acts are very difficult to justify in Darwinian terms. They make no biological sense, they are beyond ordinary comprehension and appear to have no counterpart in natural history, but they touch some human nerve because they play a powerful part in myth and drama as "The Medea Syndrome."

She, you may remember, was grievously wronged by Jason,

who abandoned her for a younger woman. But Medea, unlike Susan Smith, was no helpless victim, not troubled by low self-esteem, and she rejected the option of suicide with the words: "It's not the dying I dread, but the thought of leaving my enemies alive and laughing at my corpse."[122] And she took her revenge by wiping out not only the entire royal house of Corinth, but also her own two sons, whom she butchered with a knife. With that act alone, taken without the biological pressures that can make some infanticides appropriate or at least understandable, Medea overcomes one of our most basic instincts, one the genes insist on more than any other, and becomes something unnatural. A mistress of the black arts. Heroic perhaps, but very definitely evil, by any definition.

There is only one other homicidal pattern that rivals Medea for apparent biological futility, but is far more common and even popular in the sense that it seems somehow appropriate, whatever the cost—and that is revenge. An eye for an eye, a tooth for a tooth, evening the score. It is a practice that has been condemned by the best tacticians of our time. Winston Churchill said of it: "Nothing is more costly, nothing is more sterile, than vengeance."[70] And yet it seems to have been a feature of almost every culture ever studied, in many of which it assumes the status of a sacred obligation. A life must be paid for by a life, even if it takes generations.

At a functional level, revenge can be seen as delayed retaliation. Like a fight in which the blows are separated not by seconds, but by months or years. This at least makes some sense. A counterpunch can be an effective deterrent, a warning that no attack will go unpunished, usually by a retaliation in kind of at least equal severity. For a species with a long memory, consciousness of unavoidable counteraggression at some future time, perhaps when you are least expecting it, could serve very well to keep the peace. Viewed in this light, the explicit tradition of an eye for an eye comes to look more like a strategy designed to contain, rather than encourage, revenge. It works that way among the Asmat at least, who are well schooled in the ideas of equity and equilibrium. But in some societies, blood-feuding quickly gets out of hand as individuals or kin groups use retaliation as an excuse to go more than one-up, to seize an advantage or to escalate a quarrel in the vain hope of a "final solution."

Blood feuds are usually male affairs. But a recent study in rural Turkey shows how central a role women play in keeping old hatreds alive. The bloody shirt of a slain family member is carefully preserved by a matriarch and shown repeatedly to all young male children, who are told of the cowardice and ungodliness of the assailants, who don't deserve to live. The boys are raised with the idea of revenge, told that the soul of the victim cannot rest in peace until he has been avenged several times over, and they are reminded of their duty to do so every time they see the old woman dressed in the black she has sworn to wear until such a day of atonement.[210]

Just across the Andana Sea, in Cyprus, Turkish and Greek villages turn such intercommunal killings into international incidents by reviving memories so old that they give life and legitimacy to ancient claims on that disputed territory. This is how stony ground becomes "sacred" territory where armies of dead heroes still march; how burial grounds are dug up at regular intervals, their contents scattered and gravestones defaced; and how it becomes possible for a Greek boy with a machine gun to burst into a house in retaliation for the death of a friend and slaughter a whole family of seven, including women and children, and then dismiss his excessive reaction with a shrug and the offhand comment: "But they were *Turks!*"[244]

This is how revenge spirals into mad blood-lust that spans generations and leaves both feuding sides weakened to the point of exhaustion. Gang wars in our inner cities sometimes go the same crazy way because they lack any tradition of useful arbitration—they haven't learned how to play the game right. They have no "Man Who Can Be Seen" to intercede. But it need not be this way. There is always Tit For Tat.

Tit For Tat, remember, dictates that you start nice, but then always match your opponent's last move. In other words, retaliate precisely in kind. No more and no less. This is a strategy that neither provokes suicidal escalation of the kind one sees in blood feuds, nor does it advertise the sort of weakness which would invite exploitation. In the long term, an eye for an eye works because it means just that. If they take an eye, you take an eye; if they turn a blind eye, you reciprocate by being equally forgiving. Both nice and nasty are instantly acknowledged and rewarded in kind, and everyone knows exactly where they stand.

In 1979 Robert Axelrod discovered that this strategy, and as far as we know only this strategy, is stable and resists all invasions by rival programs.[21] And that was a vital find. It showed that measured retaliation had a positive and cumulative effect that always ended in greater cooperation. And it raised the fascinating possibility that we social creatures, whose big brains have been nourished and excited by the cut and thrust of getting along with others of our kind, might intuitively appreciate the cost benefit of icy retaliation. The Italians have always said that revenge is a dish best eaten cold. That may be exactly right. Perhaps our fondness for the idea of just retribution is a sound one, based on long evolutionary experience. It is possible that the satisfaction it gives us, the spiritual fulfilment it seems to bring, despite the pain, is an adaptive response. Revenge makes sense, it feels right, simply because at some level we recognize that it also brings long-term stability to competitive rivalries.

An eye for an eye? By all means, because no feud need be interminable and many in tribal communities are indeed resolved after a single retaliatory homicide, coupled usually with a settlement of some kind. Blood money, because life in most societies does have a price. In prewar Somalia, the accidental death of a man was worth a hundred camels, that of a woman fifty.[78] Who knows to what pitiful level the crazed "technicals" have reduced it now? Among the Sebei of Uganda, they exercised a finer appreciation, asking ten cows for a young woman of marriageable age, and five cows for an older woman no longer capable of bearing children, but known to be a good cook.[150] But in our society, we have no such possibility of redress, because we are far more ambivalent about our precise worth.

Measured as a source of chemical compounds, each of us is worth about £4.99; as a source of organs for transplant, perhaps as much as £200,000. The average life is insured for only £50,000, but Bruce Springsteen's vocal cords alone have been covered for more than £3 million. Most natural pregnancies are free, but the cost of artificially creating a new life varies from about £30 for a sperm donation to over £500,000 for the average six cycles of *in vitro* fertilization.[283] Such calculations become relevant when talk turns to criminals being "made to pay" for their crime. If the death sentence is no longer an option for murder, then what does constitute proper compensation? Juries may bring in recommendations for huge sums in punitive damages, but only after years of protracted and expensive litigation in a civil suit. By which time it is possible that the killer may even have been freed from prison as a result of deals behind the scenes, over which the victim's family has no control. A typical sentence for manslaughter in the United States now is three to five years, with parole possible after just eighteen months. With the result that justice very often is *not* seen to be done and relatives still carry hate and feel an unrequited thirst for revenge.

We ignore such passions at our peril, and ought to think twice about dismissing revenge as "sterile" or somehow unseemly. We lost something of value, a pattern that made evolutionary sense, when we decided to make justice remote and impersonal. Perhaps we should think again and restore once more the old notion of "just deserts," of measured retribution that is satisfying and comprehensible to all concerned, because it is totally appropriate to the crime.

Such justice would need to differ from culture to culture, as befits the circumstances and the perception of the crime. People are far readier to resort to violence in some societies, very slow to anger in others. But even among those who frown on aggression as antisocial, there may be mechanisms which legitimate occasional outbursts. In Southeast Asia, running *amok* is such a cultural construct which still surfaces from time to time. It usually involves a young man who goes on a rampage of random murder, traditionally with a blade, killing anyone unlucky enough to stand in his way. I have seen a Malay in such a state, armed with a *kris* and apparently entranced, who cut his way through a crowd in a small town marketplace, leaving dead and dying in his wake; and who carried on slashing at will until several bystanders succeeded in stopping him by the simple expedient of beating him to death with their sticks.

No one blamed the killer or congratulated those who ended his spree. It was simply accepted that he was a casualty of the sort of desperation that society can sometimes produce, and though his victims were unfortunate, the "amok" himself was considered to have resolved his dilemma honorably. The whole affair was over in a matter of minutes, culturally assimilated and closed without any of the agonizing and heart-searching that follow similar mass murders in our society. I was able to make a direct comparison, because soon after I returned from Asia in mid 1966, an ex-Marine called Charles Whitman died in the tower of the University of Texas at Austin after killing fourteen people and wounding thirty-one others when he decided: "I can't stand the pressures on me. I'm going to fight it out alone."[40]

There seems little to separate the Malay from the Marine except that one had a revolver, two pistols, three rifles and a shotgun; while all the other had was his family's sacred blade. Both killed in a disturbed state of mind. Whitman was seeing a psychiatrist about his hostility and actually described a recurring vision he had of "going up on the tower with a deer rifle and shooting people." All anyone knew of the Malay was that he lived alone and had some gambling debts. Neither's action can be condoned, but if evil is indeed to be measured by the degree to which it disturbs or disrupts the local ecology, then Whitman was certainly the more evil of the two. The cultural environment of the Malay village allowed its equally lethal episode to be absorbed and accepted in a very graceful way.

That matters. I find that now, almost thirty years later, and even though I was there to see the blood shed by the "amok" at first hand, I somehow feel less disturbed by his passionate act than I do reading again about Whitman's cold-blooded execution of everyone who came within range of his telescopic sights. There is a difference of context—and this, yet again, is something we need to consider in any evaluation of good or evil anywhere.

Homicide rates vary, not just from country to country, but from time to time within each nation-state. We hear a great deal about violence and its prevalence among particular groups or subcultures, but the truth is that it tends everywhere to be most common among poor young men with dismal prospects, whatever their ethnic or racial background. They have good reason, if not moral justification, for trying to raise their game to compete with those who seem to have an unfair share of local resources. And one of the strange patterns in such violence suggests that escalation may be connected with the results of some of our favorite "substitutes for war," the sports of prize-fighting and football.

There are more than sixty homicides every day in the United States, and one study of these shows that there is a significant increase in the few days after a well-publicized and widely viewed heavyweight championship boxing match. And if the contest takes place between a black and a white fighter, the additional deaths tend to be of the same race as the loser.[301] This effect, though controversial, seems to be borne out by a follow-up study on the equally physical and aggressive performance of American professional football.

On the days following a major playoff game, there are always more homicides than on the days preceding the game, and these additional deaths invariably take place in the community which houses and supports the losing team.[412] The catalyst for the violence that leads to such homicides seems not to be the aggressive performance itself. There is no support here for the theory that violence breeds violence. But there is every reason to believe that the result is what matters. Losing an aggressive encounter means a loss of face, and this disappointment touches not just the performers, but those who support and identify with them. The winners go home in high spirits, full of well-being; the losers vandalize their surroundings, pick fights with each other, and beat their wives and children.

And yet the value of sport as a substitute for war seems undeniable. World Cup soccer matches may lead to the summary execution of losing players; but the level of actual violence as measured by homicides in these sport-crazy days has never been lower. The English murder rate has fallen consistently since the Middle Ages, despite declamations of social disintegration and doom by commentators in each passing century.[32] It may seem that we live in violent times, but even the famously gentle Bushmen of the

Kalahari have a homicide rate that eclipses those of the most notorious American cities.[240] All appearances to the contrary, we who live in today's industrial societies stand a better chance of dying peacefully in our beds than any of our predecessors anywhere.

Natural selection determines the way we are, pushing us always in the direction of effective reproduction and enthusiastic nepotism. The rules are simple: "Go Forth and Multiply"; and "Look After Your Own." And the things we do, or seem to choose to do in our own self-interest, tend also to be appropriate in the sense that they have evolved to maximize our fitness to survive in a traditional environment. We have now moved some distance away from that historical habitat, but we still favor such strategies and put them into play when we feel ourselves in conflict with others competing for the same resources. We become assertive, often downright aggressive, and we will kill if that is what it takes to stay fit, although we have powerful constraints against killing genetic relatives. Even these, however, can be overcome if there seems to be a higher good, a greater chance of reproductive success that can only be achieved by sacrificing close kin. It seems, therefore, that we have no taboo, no absolute prohibition, as far as killing is concerned. It is something our species does when it has to, sometimes even with what looks like pleasure—or is it perhaps just curiosity?

Something happened in England on February 12, 1993 which, for once, justified newspaper hyperbole and earned its right to be known as "the story that shocked the world." The facts are simple. At exactly 3:38 P.M. on that day, James Bulger, a child not quite three years old, wandered away from his mother in a shopping precinct outside Liverpool in Lancashire. He was recorded three minutes later by a security camera on another floor in the company of two older children. A taxi driver and a woman shopper saw two boys drag the child across the road outside the center a little later, and a further thirty-six witnesses watched, but didn't interfere—assuming that the three were related—as the boys took James, who seemed disturbed, on a two-and-a-half-mile walk down a well-traveled bus route, past a roundabout, into a do-it-yourself shop and a pet store, out on to the verge of a railway line. There, two hours after kidnapping him, they tortured James to death. They did so deliberately, hitting him with at least twenty-seven bricks and a cast-iron bar weighing twenty-two pounds, causing forty visible wounds and multiple fractures of his skull. Then they took off his trousers, covered parts of his face and body with blue paint, apparently stolen en route for this purpose, and finally laid him on the railway track so that the next train cut what was left of him in two. Then the two boys, both ten years old, dressed still in their school uniforms, visited a video store as if nothing had happened.

Jon Venables and Robert Thompson were arrested less than a week later and charged with murder. There was never any question of their guilt. The two had left a trial of damning evidence, not even troubling to remove James's blood from their shoes or traces of the blue paint from their jackets. Under questioning, they blamed each other, but neither then, nor at any point in their sensational trial, did either of them give a clear explanation about why they might have done such a thing. Others tried . . .

Everyone in Britain felt personally involved. How could such a thing have happened? Who was to blame? Could it have been prevented? Might it happen again? Is society itself at fault? The agonized soul-searching went on for months over what the *Independent* called "the worst murder of our time."[11] The people of Merseyside, already in the depths of depression, struggled to find a way of living with the guilt of such horror in their midst, and were described as being "consumed in the bonfire of misery."[177] The British government and the Church of England clashed over responsibility in high places and, as the shock wave of revulsion spread, newspapers all over the world wondered: "Could it happen here?"[194]

The experts, of course, gathered in a blizzard of theory to talk of "moral degeneration and decline"; of "a society whirling savagely out of control";[302] of "the dangerous psychological process of the dyad"—a group of two who jointly commit acts neither would individually countenance;[300] of "the death of innocence," and "continuums of behavior, from children pulling wings off butterflies at one end, to sadism at the other."[33] Everyone struggled to find meaning in the tragedy, but to the very end it proved elusive and refused to be used in quite that way.

The judge who heard the case and pronounced the verdict at the end described the death of James Bulger as "an act of unparalleled evil and barbarity" and the conduct of the two killers as "cunning and very wicked."[303] And the police detective superintendent responsible for finding the culprits characterized them as: "wicked beyond any expectations," with "a high degree of cunning and evil."[49] His team is still haunted by what one of them describes as the "terrible, chilling smile" that Thompson exchanged with Venables at one point during the trial. "It was a cold smile—an evil smile . . . which said they knew they were responsible, and thought they were going to get away with it."[405] Too young to be sentenced and committed to prison, Jon Venables and Robert Thompson have been detained "at Her Majesty's pleasure" in separate high-security schools with small groups of other dangerous children serving what amount to life sentences. These two are, however, in a different class from the others, locked away as "the most perplexing specimens in British captivity."[128]

Large crowds gathered at every stage of the trial, shouting "Kill the bastards" and "A life for a life." It seemed clear to those who watched this anguish and heard such imprecations that, given the opportunity, these ordinary God-fearing people could have lynched the boys. Out of a desire for vengeance and pure justice perhaps, but also from terrible frustration and a sinking feeling that it could have been their child that was killed or, worse still, their children doing the killing.

We seem to be faced, in such circumstances, with a dilemma so close and yet so incomprehensible that it produces a kind of moral panic. At the core of this killing is the awful fact that it was carried out by children against an even younger child. For that transgression, no remedial sentence seems enough. That puts us all at risk. That makes every childish prank, each tiny act of social rebellion, seem like the thin edge of a terrifying wedge that leads inexorably to the worst and the darkest deeds anyone can do. And we don't want to believe that the veneer of civilization can be that thin.[11]

It probably is. William Golding in *Lord of the Flies* suggests as much, letting his stranded children revert very quickly to savagery once they are freed from the watchful restraint of adults.[149] Richard Hughes in *A High Wind in Jamaica* tells the story of a child who killed, and of seeing her later standing among other little girls at her new school in England: "Looking at that gentle, happy throng of clear innocent faces and soft graceful limbs, listening to the careless, artless babble of chatter rising, perhaps God could have picked out from among them which was Emily: but I am sure that I could not."[200]

There may well be a vicious streak that lies just beneath what we imagine to be innocence. But two-year-old, blond-haired, blue-eyed James Bulger was innocent, wasn't he? Perhaps. And perhaps he could be so only because he was the product of a united family whose strength became very evident after his death. Ian McEwan, whose fiction bristles with images that are replete with menace, is convinced that the loss of innocence is a theme that has preoccupied us since our mythical expulsion from the Garden of Eden. "The violent child is the most potent image of violated innocence we have," he says. "If humanity is capable of this, then perhaps we are beyond redemption."[33]

The children who killed James Bulger were the products of poverty, broken homes and abusive parents. They were persistent truants, allies in underachievement, at odds with their own society even before the murder. But the same is true of hundreds of thousands of disadvantaged kids who don't hang around shopping malls looking for someone to kill. It emerged during the trial that Thompson and Venables, on the same day they succeeded in kidnapping James, tried at least once to lure another child away from its mother. They were looking for trouble and had something specific in mind.

There is a chilling coincidence in that the last video rented by Jon Venables's father before the Bulger tragedy was *Child's Play III*, a story about a demonic doll called Chucky who wears dungarees, has blue paint splashed on his face, abducts a boy and tries to kill him under the wheels of the ghost train in a fairground called the Devil's Lair. There is no evidence that either of the boys ever saw this video, but a further alarming note is provided by the fact that Venables, in his attorney's words: "has convinced himself he played no part in the killing. He talks about it as if it was something he had seen."[286]

The corrupting effects of cinematic violence and "video nasties" are an all-too-easy target. Anthony Burgess grew tired of having to defend *A Clockwork Orange* against attacks that the book, and Stanley Kubrick's film of it, encouraged violence.[51] "Neither cinema nor literature can be blamed for the manifestations of original sin. A man who kills his uncle cannot justifiably blame a performance of *Hamlet.* On the other hand, if literature is to be held responsible for mayhem and murder, then the most damnable book of them all is the Bible."[52] The film of *A Clockwork Orange* is still banned in Britain, and the book has lost its edge. After committing robbery, rape, torture and murder, the lead character says: "That was everything. I'd done the lot, now. And me still only fifteen." In 1962 that was a shocking disclosure. Not now. Try "ten."

The real issue is what makes images that are violent so resonant; what is it about ourselves that lays us open to such seeds of evil? There is no easy answer. The minds of Jon Venables and Robert Thompson seem already to be beyond our reach. But there is something Thompson said in his testimony that rings a warning bell. When first accused of killing James Bulger, Thompson became indignant and dismissed the suggestion with a defiant: "If I wanted to kill a baby, I'd kill my own, wouldn't I?" Most commentators on the case have passed over this strange suggestion as a pathetic attempt at reasoned denial, but one heard something very different and more sinister there. Something that makes sense only when you know that Thompson was the fifth of six unruly boys in a family whose role model was a drunken bully of a father, who left when Robert was five. And seven months before the murder, his mother produced a seventh son, this time by another man, saying: "I don't care about the other boys. This one is going to be different, better."[358]

Piers Paul Read, with a novelist's ear for nuance, heard in that comment an admission that Robert was jealous of the new baby and that the idea of killing it had crossed his mind. Nothing too surprising in that. It is common for children to resent new arrivals and this one, remember, was a half-brother with whom he shared only one-quarter of his genes. That twenty-five percent, however, may have been enough to protect the child, even from Thompson. The six brothers had become used to fending for themselves and closing ranks against outsiders with the fatalistic comment: "Yeah, well . . . it's always our family that gets the blame." But, suggests Read, "It was only to be expected that the ten-year-old Thompson should feel angry and aggrieved. And just as under analysis a patient can transfer his feelings for a parent on to the analyst, so in the mind of Thompson the transference was made between a rival for his mother's affection and a toddler picked at random outside a butcher's shop . . . If we accept as the jury did that Thompson was the dominant one of the pair, and that the acts he ascribed to Venables were, in fact, his own, then the motive becomes clear. The reason given by Venables, according to Thompson, for killing James was "because he felt like it." Undoubtedly that was the reason. Thompson tortured and murdered the child because he felt like it, and he felt like it because he loathed his sibling.[313]

That makes biological sense. A confusion between hate and loyalty, compounded and encouraged by genetic distance, could easily have flowed over into a violent act against a complete outsider. The details of the act may in their turn have been drawn, consciously or unconsciously, from a nasty video, but Read quite rightly turns most of the moral panic in Britain on its head by suggesting that the question posed by the murder of James is not: "Why did it happen?" but: "Why does it not happen more often?"

It probably will. On average there is one child under the age of five murdered each year in Britain, and killings by children, though rare, are not unknown. There have been six in the last twenty-five years, most notably eleven-year-old Mary Bell who, in 1968, was convicted of manslaughter for strangling two boys aged three and four in Newcastle. Like Thompson and Venables, she was described as "evil," a "monster," a "little fiend" and a "bad seed." Like Thompson, she was bright and quick, and ruthless. Four days after her first victim's death, Mary knocked on the door of his home and, smiling, asked to see him. His mother said: "No, pet, Martin is dead." "Oh, I know," said Mary, still smiling, "I wanted to see him in his coffin."[344] It was Bobby Thompson who, five days after killing James Bulger, bought a single red rose and placed it on the railway embankment, with all the other floral tributes to the dead child. Like Thompson, Mary Bell was arrogant and unrepentant, aware that she was in trouble, but unable to understand what all the fuss was about. And like Thompson, she was an energetic child with a crucial gap in her life.

Both are the product of homes so badly broken that the one remaining parent can only be classified as dysfunctional. Mary's mother was a disturbed unmarried teenager, just seventeen when her child was born, unable to feed or clothe her properly, leaving Mary alone or even giving her away to strangers. Robert's mother was a heavy drinker, a pub brawler, who abused her children or let them be taken into care. Neither he nor Mary had any kind of normal affection, training or discipline in their early years and both became inhumanly cruel. Thompson tied cats to railway lines, Bell pinched and pricked babies for fun: "Because I like to hurt." The abuse of animals or babies is a very common part of the profile or case history of those who end up killing. A warning that something has prevented the normal development of a moral sense, of an ability to tell right from wrong. What is missing, in short, is the "knowledge of good or evil."[424]

The evolutionary origins of morality remain obscure. Behavior patterns come into being largely because they benefit the individuals involved, or at least provide sufficient benefit for their immediate genetic relatives. It is difficult, in biological terms, to imagine how anything can be selected to provide benefit to the species as a whole unless it first offers such clear advantages to the individual. Being moral, behaving well, is not obviously such a pattern. And yet animal behavior is peppered with examples of something very like social disapproval.

A jackdaw that intrudes on another's nest is attacked by all members of the colony.[245] A goose with damaged plumage, or a bear wearing a radio harness, will often be harassed by group members simply because it looks different. A child going to a new school very quickly learns to dress and speak like others there, simply to avoid being conspicuous. There is in conformity of this kind the beginnings of a code, obedience to which simply avoids the possibility of punishment. But a far more powerful motive to conform lies in being persistently rewarded for doing so. Sometimes it may be reward enough simply to be accepted by a society that provides greater protection from predators, but there are many less direct benefits, such as social grooming, which can reinforce togetherness. And there is every evidence that individuals learn about such benefits, and how to earn them, at an early and critical age.

Primatologist Hans Kummer tells of a mother dog that consistently prevented her newborn pups from going into a part of the garden that she herself had been trained not to enter.[235] Children frequently spank their dolls for behavior for which they have been disciplined.[318] We all seem to learn from, and to teach each other about, how the world works; and we begin to do this almost as soon as we are born.

Even seven-week-old human babies seem to understand and enjoy the principle of the dialogue, taking turns in communicating with an adult by an alternation of cooing sounds.[334] And as they pass their first birthday, children commonly show and give objects to other people on their own initiative, apparently willing to share without prompting, direction or praise.[317] Such behavior, of course, is self-rewarding in the sense that it elicits response and engagement in play; but these and other "prosocial" acts, such as friendliness and affection, all play a vital part in building up an individual's tendency to do some things rather than others. To help rather than hurt, to cooperate rather than compete. Even as infants, we are active partners in interactions with others. We learn how to make reliable predictions about their behavior in response to ours. In a very real sense, we teach our parents how to cope with us, but if parental care and response are lacking, or even missing altogether at this crucial time, some vital connections never get made.[316]

Richard Dawkins, in exploring the nature of the selfish gene, made what I believe to be one of the most radical and useful suggestions in the history of ideas. It is one which concerns ideas themselves as the carriers of human culture. Suppose, he says, that we have given birth to a new and more rapid kind of evolution that involves culture rather than chemicals; then the process will need pieces on which to operate. Genetics has genes, so culture must have its own units of transmission—and Dawkins calls these *memes.*[101]

Memes are tunes, catchy phrases, fashions, ways of making pots or playing games. "Just as genes propagate themselves in the gene pool by leaping from body to body . . . so memes propagate themselves in the meme pool by leaping from brain to brain." Viewed in this light, memes become living structures capable of implanting themselves in another mind like viruses which parasitize the genetic mechanism of a host cell. And, like viruses, they are forced to compete with one another in a truly Darwinian fashion, struggling for access to the most successful minds and bodies. To do this, they need to demonstrate that they have survival value in the new environment provided by human culture.

The meme as an idea in its own right has proved highly successful. One of those ideas which look like evolutionary preadaptations, things that seem to have been just waiting for their proper time.This one appeals directly to human minds because, like a Goldilocks solution, it feels "just right." It fits right in. Dawkins sees it acting on a very broad front, even suggesting that religion is a "virus of the mind," a cultural analogue of catching flu or yellow fever. Something, perhaps, which happens to you when you are young enough not to know better; or when, as an older individual, your rational defences are lowered by some other factor that reduces your resistance to infection.[102]

Anthropologist Mark Ridley, noting that religion of one kind or another is common to almost all human cultures, suggests that human brains must have evolved in ways that favor imposing symbolic meaning on the external world, and that it was only much later that a minority of our species, including Richard Dawkins, became infected with the newer mutation of rational scientific thought.[320] When religious conversion, the spread of the religion meme, was practiced largely by force in crusades and holy wars, then rationality never stood a chance. Religious enthusiasm gave land a symbolic value far above its actual value as a resource; and those in a fever of infection by the religion meme took such irrational risks in battle—secure in the belief of their reward after death—that rationality was easily overwhelmed and outcompeted.

As it still very often is. The Gulf War of 1990 inflicted a punishing military defeat on Saddam Hussein, but his refusal to concede the reality of the catastrophe, his denial of spiritual defeat, robbed the Allied coalition of much of its political point. Iraq may have lost the war in the air, but the idea of a militant Islamic faith is still very much up there, competing successfully—a meme of undiminished potency.

Arguments about religion seldom satisfy, because rationality is hamstrung in the end by the fact that objective evidence for anything at all is rather poor, and religious believers routinely go well beyond the evidence anyway, turning ignorance into a virtue by making belief "an act of faith." But Dawkins's idea of infection is an exciting one. It begins to make sense of what and how we learn. Memes, like genes, can mutate. In fact they change so fast that no exact replication is ever possible. Each mind tends to play the same tune a little differently. That is one of our extraordinary strengths, but it seems to me that the best chance of useful replication for a meme will be when the recipient is young. Just as newly hatched ducks can become imprinted on any moving object, or chicks entrained to follow a flashing light, so we too pass through a highly sensitive period when we suck up information, any information, like blotting paper. We seem to be designed to do that.

Studies on the origin of language, for instance, show that human infants can discriminate successfully between even the most difficult phonemes. Each language uses sounds a little differently, but any baby under the age of six months can tell whether it is hearing fragments of Czech, Hindi, or Inslekampx (an Eskimo tongue); something its parents cannot do, even with 500 training trials or a year of university courses in those languages.[233] Babies, it seems, come into the world ready to do impossible things, primed to pick up the intricacies of any of the 6,000 tongues now in use somewhere on the planet—as long as these are presented to them in time. Wait too long, as most of us have, and you are lucky to have a working knowledge of more than one. The window of such opportunity is open wide, but not for very long. Our chances of infection are good, but only while the offer lasts. I suggest that the window for picking up the kind of habits that make us moral may be just as narrow and just as critical.

Take it and you pick up a vital infection—the seed of morality. It may not matter exactly what you get, as long as you get a decent mix—a little like those cocktail vaccinations that were once used to protect us from childhood diseases. All that seems to be necessary is that you take on sufficient outside material to mix with what the genes already provide as general human inheritance. Enough to provide a baseline, a basis for comparison, and to set up a sort of "antibody" reaction system that will guard against later, more malevolent, infection. Just as sex provides a vaccination that protects us against parasites, so knowledge seems to vaccinate us against the kind of imbalance and disequilibrium I have been calling evil. Get the shot and you stand a decent chance of getting by. Miss it and you never make it, no matter how hard you try.

Robert Thompson and Mary Bell missed it. Not as a result of truancy, though both children practiced that enough to become distant from the world they nominally inhabited. The damage to the two was done by the time they became old enough to go to school. At five, Thompson was already pulling the heads off live baby pigeons. To dismiss them as "evil" is too easy and merely begs the question of how they became that way.

Parts of the British press were very ready to label James Bulger's killers as "depraved young monsters" and to declare that "original evil exists."[362] It is a comfortable position, reminiscent of the sense that poet and literary critic Samuel Taylor Coleridge made of Iago's role in bringing down Othello. He described Iago's famous final soliloquy as nothing more than the "motive-hunting of motiveless malignity."[76] Meaning that Iago had no motives, though he looked for them in order to justify his actions to himself. He was merely malign, totally evil and therefore beyond the need for justification or explanation. It was feelings of that kind that turned a crowd of ordinary people in Lancashire into a mob howling for blood at the police vans as they brought the ten-year-old "devils" to court. The need for vengeance is powerful, but it needs to be tempered by the knowledge that it could have been our children coming home with blood on their shoes.

Novelist Martin Amis, himself the father of a ten-year-old, had this in mind when he rented a copy of *Child's Play III* and sat down to watch the story of the doll called Chucky that comes to life and starts killing people. Afterward, he said: "I felt no urge or prompting to go out and kill somebody. And I know why too. It's nothing to boast about, but there is too much going on in my head for Chucky to gain sway in there . . . What we have to imagine is a mind that, on exposure to Chucky, is already brimful of Chucky and things like Chucky. Then, even if you mix in psychopathology, stupidity and moral deformation, Chucky is unlikely to affect anything but the *style* of your subsequent atrocities. Murderers have to have something to haunt them; they need their internal pandemonium. A century ago, it might have been the Devil. Now it's Chucky."[7]

He's right, of course. But as a fifteen-year-old in Birmingham points out: "Thousands of children here live in broken homes and watch these violent films, but don't kill."[209] And that too is true. But where Thompson and Bell differ from such sane and sensible points of view is that they have nothing against which to measure their behavior. When we become programmed with the information we will need to make moral judgments later in life, we follow just one rule: "Believe what you are told." There is no time for anything else in those hectic days when the window is open wide. You take what you can get and sort it out later, discarding in hindsight the bits that don't make sense, incorporating the bits that seem to help to make plans that work, strategies that get you where you want to be. We can make such value judgments because we have a moral yardstick, something to work with, even if it's only a bad example drawn from the Old Testament or a wayward uncle. Thompson and Bell, it seems to me, were operating in a moral vacuum, making it up as they went along, looking for something they wouldn't recognize even if they found it, lashing out in the anger and frustration of not knowing which way to turn. This is not an excuse for what they did, merely the semblance of a reason for the way they behaved.

I wasn't convinced by reports of "evil" from that imposing room in Preston Crown Court, and managed to borrow a press pass for part of one of those seventeen strangely sunny November days in 1993 when the two suspects, known then only as Boy A and Boy B, were on trial. The two defendants sat in the dock flanked by social workers, wearing jackets and ties, their hair neatly cut. Both were then eleven years old, one a little chubby and disinterested, the other given to toying with his tie, letting it unroll, again and again, down his chest. There was something about chubby child A, whom I now know to be Robert Thompson, that made it clear he would have been the dominant one, the one who always took the lead. But there was nothing else. No horns, no tail, no mark of Cain—and that worried me.

Killing in cold blood is an unnatural and dreadful thing to do. It ought to leave its mark on someone who has taken part. There ought to be a sign. Something that we—whose big brains developed as a direct result of the demands of social life, whose whole recent evolution has been predicated on a superior ability to read, and make intelligent use of, information about each other's state of mind—should be able to detect. You would think that would be a vital prerequisite for a naked ape, otherwise unarmed, who had to live by its wits. Something with real and obvious survival value. If we have a talent for detecting those who cheat on us, who fail to play the altruistic game by reciprocating in kind, you imagine that we would have evolved, and nature would have selected for, an ability to detect whether someone had blood on their hands. But after three hours in that courtroom, I was disappointed. I had to agree with Robert Hughes's observation about his fictional Emily. There seemed to be nothing about either boy that would have made it possible for me to pick them out from a lineup of their classmates at St. Mary's School in Walton; nothing that would separate them from all the others there who didn't kidnap and kill a child. Thompson and Venables looked like very ordinary children—and that, in itself, was very scary.

The novelist Will Self makes the same point, with feeling, about a multiple murderer who happened to interview him at the state employment bureau in London: "It is now almost aphoristic to say of evil that it is banal, chillingly ordinary. But the fact remains that the Dennis Nilsen I met seemed to me, as he did to his colleagues and acquaintances, to have had no charge, no resonance, no aura about him whatsoever. His chain of fifteen murders, with all their attendant necrophilia and dismemberments and his hoarding of body parts, had yet to be discovered."[343]

Journalist Anne Schwartz had similar difficulty in accepting the apparent normality of another serial killer: "Many people came to court to see if they could sense any underlying evil in Jeffrey Dahmer by looking at him. I remember being filled with a strange sense of anticipation as I waited in the courtroom to see what a serial killer looked like. I had thought the pupils of his eyes might do spirals. What he had done was awful, but I could not get over how ordinary he looked. The times I saw him up close, I saw nothing there. He did not appear crazed, like mass murderer Charles Manson, nor did he exude the charm of serial killer Ted Bundy. There was nothing to him."[340]

I know the feeling. It is surprising. And she is right to have wondered about his eyes. When we look at one another, all we see is death. Dead skin, dead hair, dead nails—all pieces of the home-grown suit of armor we are obliged to wear to protect our living tissue from the ravages of oxygen. The only glimpse of life we get in one another lies in the tinted cells of the iris which show through the transparent corneas of the eye, those "windows of the soul." Small wonder that we pull our lower eyelids down in a global gesture which says: "Watch out. Be alert. There's someone in here, and I've got my eye on you!" "I don't have a wooden eye," they say in Germany when they want you to know that you are being observed.[279] And, as predominantly visual animals, it is no surprise that having another's eyes on you, being stared at directly, should have come to be so threatening that many of us feel that we can sense when someone is doing so behind our backs.[351] Such feelings may be justified. Recent laboratory experiments show that, significantly more often than not, someone *was* staring at the time.[423]

The word "fascination" comes from the Latin *fascinare,* which referred to an enchantment or a spell cast by a look—usually a bad look, hence "the evil eye." This belief in transmission of influence *from* the eye is remarkably widespread, covering all of Europe, most of the Arab world, Southeast Asia and Central America, in all of which it is usually malevolent. The loss of an eye in any of these areas seems to be seen as a disturbance of social order. Perhaps the man with the black eye-patch is hiding more than an empty socket?[142] And the eye of god—whether this be Ra, Atum, Osiris, Horus or Jehovah—is everywhere also seen as a remedial talisman, worn as an amulet, or painted on the prow of a boat, or on the bumper of a car. Few other beliefs have ever been so widespread or so continuous, and there is no other body part that so directly represents, and is seen to be so active in, the forces of good and evil.[207]

Perhaps with reason. We routinely damp down overstimulation by taking the quickest available way out of a stressful situation. We close our eyes. We lower the shutters on the mind. And we learn to be wary of those who cannot face us directly, but look away with "evasive" or "shifty" eyes; and feel disconcerted by those who send conflicting signals with eyes that "stutter" or "stammer" by closing at inappropriate moments, cutting the person behind them off from us in ways over which we have no control. What, we wonder, have they got to hide?[278]

Writer Alec Wilkinson begins his searching interview of John Wayne Gacy, who sexually abused and murdered at least thirty-three boys, with a strictly optical analysis: "His eyes are small and remote and measuring . . . He has concealed the complexity of his character so assiduously that a person is left to imagine the part of him that carried out the murders."[416] Before double murderer Gary Gilmore was executed by firing squad in Utah in 1977, Norman Mailer spun his reportage around the black hole of the killer's mind so dexterously, and so visually, that his Pulitzer Prize–winning book *The Executioner's Song* was also turned into a rock song called "Gary Gilmore's Eyes."[257]

Something similar, but collective, appears to have happened more recently in Africa, where: "It is said that many of the Hutu tribespeople who crossed the Rwandan border into Tanzania by the thousands in a massive wave of refugees, had 'a killer in the eye.' That was the same maddened, pitiless look they had when they raised their machetes against men, women and children of the minority Tutsi tribe and even against moderate Hutus who sought to protect them or were associated with them, and sliced open their arms and chests, stuck blades into their mouths and genitals and lopped off their heads. It was the same look evident on the faces of those who did not commit such acts but nonetheless stood by and cheered the genocide on . . . Even now, in the dim, smoke and squalor of the refugee camps, a hardened glare will show itself from out of a tent or a crowd, sufficient to remind one of all that is murdering Rwanda."[325]

The eyes have it. And what they ought to have, what you expect to find there, is evidence of "Satanic" powers, or at least of dark, atavistic, primal forces. But with the exception of permanently crazed individuals like Manson, you seldom do. What you get is an eleven-year-old like Robert Thompson who seems totally unmoved, even when he listens in court to his own voice, coming out of the speakers, recounting the last moments of James Bulger's life, stunning everyone else in the room into a sort of sick silence as the next witness links Thompson's bald testimony to the evidence of his shoe prints on the infant's face. Not a word was said directly to either boy while I was in the courtroom. In fact, neither was called upon at any time in the entire trial. Jon Venables spent it looking occasionally disconcerted, fluttering his hands, trying to catch his parents' eyes; but Thompson, according to David Smith's book on the killing: "revealed no sign of distress, anxiety or any other emotional reaction."[358] This inertia has been variously interpreted as "the boredom of indifference" or "the impassivity of an unfeeling psychopath." "Robert's eyes," said journalist Stephen Glover, "know nothing of contrition."[148] All that may be true, but as the boys were ushered out of the court during a recess, I saw something that still turns restlessly in my mind.

As Robert Thompson left his seat, he looked briefly at the public seats behind him, and it wasn't boredom or indifference that showed. Those are emotions, negatives ones to be sure, but nevertheless emotional states that are suggestive of response. There was nothing in Thompson's eyes. Nothing whatsoever. None of the glare that several of his police inquisitors described as "difficult to explain." It was a blank face and yet not one frozen still in shock; just blank, a mask without identity, dead skin over dead eyes. I had never seen anything quite like it before. But I have since, and I think there is a connection.

Three months before the Bulger trial, on August 2, 1993, in the small town of Savona in New York State, a thirteen-year-old boy led a four-year-old into a patch of woods, choked him into unconsciousness, battered him to death with rocks, sexually abused him with a stick and doused the child's body with red Kool-Aid from his own lunch pack. Eric Smith is a freckle-faced, red-haired boy with glasses draped over slightly folded ears. He looks like an older version of Dennis the Menace, with all that cartoon character's totally unthreatening dishevelment. He confessed to the killing of Derrick Robie six days later, describing in detail what he had done, clearly enjoying the attention of his interview with the local District Attorney. And when asked why he had killed the child, said: "I don't know. I just saw this kid, this blond kid—and I wanted to hurt him."[62]

The two boys knew each other and had played together on occasion. Enough perhaps to make Eric, who had often been teased about his looks and his speech problems, jealous of the outgoing, bright, attractive younger child in his "Party Animal" T-shirt. Nothing unusual here, just the normal and inevitable growing pains of being adolescent and insecure. Enough to hate, but not enough to kill. Eric's family is extensive, four generations of them, all living nearby. Lots of hugs and kisses and no social isolation. Eric visited frequently with his thirteen aunts and uncles and twenty-eight cousins, with whom he was invariably gentle; "like a little babysitter." An older boy with whom he went fishing said Eric "was so soft that he couldn't even bait his own hook." It was his great-grandfather who went with him to the police, and his grandfather who sat outside with his mother, while Eric confessed. Both later described him as "very loving" and found his actions totally bewildering. "My God," said the grandfather, "I should have been able to see somehow. There should have been something there to see!"

Indeed. And there is, but it is still very hard to see. Mainly because it is not something, but the lack of something—and how do you know when you are looking at nothing? It is difficult to recognize something by its absence, unless one has become conscious of such absence—like the man on the stair who "wasn't there again today." And in Savona I got a glimpse again of the absence that had worried me in Preston.

CBS television covered every stage of the year-long prosecution of Eric Smith. They called their special report on the murder: *Why Did Eric Kill?* and came to no useful conclusion, but what they did show were extended close-ups of Eric's face before and during his trial. He is a very different child from Robert Thompson. Three years older, brought up in an apparently happy home in a quiet little middle-class town. A "good boy" by every account, and yet he killed as viciously as "bad boy" Thompson, also for no apparent reason; and was "happy"—this was the word the police used to describe his attitude—to take part in the investigation. He laughed a lot. Thompson, it is said, couldn't stop laughing while kicking and battering James Bulger to death. And both he and Eric ended up with the same strange look.

It isn't the look of withdrawal. I have worked with autistic children and know that face well. This is something else. Or perhaps I really mean some thing else. The eyes are like lenses on an unmanned security system, watchful without any hint of personality. These are windows on an empty room; the occupant is out.

It has become a cliché that after every murder which turns out to have been particularly brutal, one of the suspect's neighbors will say: "But he was such a nice quiet boy. It is hard to believe he could do a thing like that!" There are always those of course who become very wise after the event and claim to have felt all along that "something was wrong." Robert Thompson put up enough warning signals to justify a major alarm, but he was after all only ten years old, and just one bad boy among many others in a difficult environment. I am aware of the danger of being smart myself with the benefit of hindsight, and admit that I went to Preston hoping to find evidence to make some sense of Bulger's death. And I am far from certain that what I think I have found in both

Preston and Savona makes any kind of sense, but I am haunted by that "look" and am beginning to see it again in film and photographs of John Wayne Gacy, Frank Potts, Henry Louis Wallace, David Berkowitz, Jeffrey Dahmer, Henry Lee Lucas and Robert C. Hansen, to name but a few adults who have among them slaughtered over 200 people. It is the look of a mind less than human.

The look I am trying to describe has something in it of one dreamed up by science fiction writer Theodore Thomas in his classic short story about astronauts who went into space and came back changed. "These were men, but the eyes were different. There was an expression not found in human eyes. It was a level-eyed expression, undeviating. It was a penetrating, probing expression, yet one laden with compassion. There was a look in those eyes of things seen from deep inside, of things seen beyond the range of normal vision. It was a far look, a compelling look, a powerful look set in the eyes of normal men."[373] If this is the look of humanity fully realized, the best that we can be, then turn it over, turn the other cheek and get a glimpse of the worst. Take away the normality and the compassion, turn the look more in than out, and instead of the fabled "far look" you have the crippled version I am talking about. A look literally without character, hope or pity. *Nosferatu.* The face of the Other.

Murder is the one crime in New York State for which a thirteen-year-old can be tried in adult criminal court. So Eric Smith was charged with second-degree murder and pleaded "Not Guilty." The jury believed he knew what he was doing and returned a verdict which carried a sentence of "nine years to life." Theoretically, he could get out by the time he is thirty.

The family of Derrick Robie, who never quite made it to his fifth birthday, were relieved by, but took little pleasure in, the verdict. It was the right conclusion for everyone in the town of Savona, providing some sort of legal sanction with an appropriate remedy that gives them permission to go on with their lives, but it didn't answer any of the questions. And Eric's extended family are left wondering exactly where they went wrong. This seems to be the inevitable fate of all genetic relatives of those who kill. They cannot help but agonize about their inheritance, and the dilemma is clearly most grave for those whose kin have killed not just once—that can be excused on a number of grounds that allow for possible redemption—but who go on to kill, again and again.[85]

Mass murder is nothing new. It has been happening ever since we armed ourselves in ways that made killing easy. But mass murder stretched out over a long period by an individual who stalks, kills and often tortures a number of victims chosen almost at random is a relatively modern phenomenon. Gilles de Rais, one of the richest men in France, was executed in the fifteenth century for killing, by his own admission, over a hundred children. Elizabeth Bathory in seventeenth-century England was convicted of personally killing some 650 young women for the peculiar purpose of bathing in their blood. In Paris, Joseph Philippe and in London, Jack the Ripper—whoever he may have been—made careers of killing prostitutes during the nineteenth century. But the habit of killing a number of innocent people in a series long enough to attract attention has made the "serial killer" a very modern aberration.[315]

It is a global problem, found on every continent except Antarctica, but it is also a pattern with a focus. Three out of four serial killers live and work in the United States, where some FBI estimates suggest they may account for over 5,000 lives a year—fifteen people every day, twenty-five percent of all known murder victims. England ranks as number two in these homocidal superstakes with seven percent, and Germany third with five percent. Nine out of every ten serial killers are men, and eight of these—even in the United States where most homicides involve black men—are white. Most serial killers are loners, about half of them working in a very limited home area where they hide in plain sight, and at least one in every five is never found.[285] The FBI believe that as many as 500 may be at large right now in the United States alone.

One they did catch in 1957 died in an asylum in Wisconsin in 1984, still largely unknown. His name, for the record, was Edward Gein and he began his psychotic career by raiding local cemeteries to bring back body parts to his home, where police later found decorative displays of human bones; human skin used for lamp shades, wastebaskets and upholstered chairs; and noses, lips, labia and nipples strung to be worn as belts and necklaces. When these resources proved insufficient to satisfy a man "ambiguous about his sexuality," Gein took to creating his own corpses, which provided him with a female scalp and face made to order, a skinned-out vest complete with breasts, and female genitalia that he strapped above his own for ceremonial occasions when he liked to dance beneath a full moon.[285]

It was these curious exploits which inspired the highly successful films *Psycho, The Texas Chainsaw Massacre* and *The Silence of the Lambs* which have, over three decades, established and encouraged an extravagant profile for the serial killer, creating a new and potent meme, perhaps even a disease, that seems to be spreading at a rate fast enough to make it menacing. There is nothing in the rules that says successful memes have to be nice. All they have to be is catchy, and this one is catching on rapidly.

Joel Norris, a psychologist and consultant on criminal cases in Georgia and Florida, has recently produced a psychobiological model of the serial killer's mind, his method and his madness, which at least puts the problem into some sort of order. According to Norris, the compulsion to kill represents the serial performance of a "morality play" in which the same story is repeated, again and again, with different victims. In telling the obsessive tale, the killer establishes elaborate rituals, setting up a "behavioral skeleton," which he wears on the outside like an insect, using this architecture to support his fantasies. And, Norris argues, because such killers seldom act rationally, but respond automatically to specific stimuli, it is possible to codify and predict their behavior.[189]

Norris identifies seven key steps in a typical serial killing:

1. The "aura" phase—in which the killer withdraws from reality and builds a fantasy that demands to be satisfied.
2. The "trolling" phase—involving an active search for suitable prey on carefully chosen hunting grounds. A time when the killer becomes very alert and focused, stalking victims like an accomplished predator.

3. The "wooing" phase—during which victims are charmed, disarmed or tricked into putting themselves at risk.

4. The "capture"—a moment that is savored on the killing ground.

5. The "murder" itself—which is usually heavily symbolic, staged to recreate a moment in the killer's own childhood, bringing an emotional high and sexual release.

6. Followed by a "totem" phase—in which the killer tries to prolong the intensity by taking photographs, dismembering, eating or preserving parts of the victim or their possessions.

7. And finally, a "depression" phase—as illusions fade and the killer realizes that nothing has changed.

During this reprise, the killers may be lucid enough to turn themselves in and confess. Though, even when they do, they are seldom believed and frequently turned loose to start all over again. But, more often, the feelings of guilt pass and the relentless cycle begins all over again, going on and on until they are eventually caught red-handed.

This sequence is persuasive and fits well with the stereotypical background of violent upbringing and negative parenting that seems to be common for most multiple murderers. Norris diagnoses serial killing as a "disease in which the traditional polarities of reward and punishment and love and hate are reversed." That too may be true, but the idea of a chain of appetitive behaviors leading to an inevitable consummatory act is somewhat old-fashioned. It is no longer seen as a simple fixed sequence even in animal behavior, and probably can't be relied on to predict far more complex human psychopathology. And it doesn't even begin to address the problem of why only a tiny majority of such "damaged" people now grow into serial killers, while the vast majority do not.

There seems to be good evidence for a genetic component in criminal behavior. Studies on twins in ten different countries show that, if one twin commits a crime, there is a fifty-two percent chance that the other twin will also become a criminal, but only if they are "identical" twins who share a hundred percent of their genes.[87] If the two are just "fraternal" twins, sharing only half of these genes, then the likelihood of the second twin following suit is reduced by more than half, to just twenty-one percent.[69] And even when twins are separated at birth and reared apart, there still seems to be some evidence that their tendency to be antisocial is at least partly inherited.[163] How this might happen remains moot, but antisocial behavior of various kinds has been linked to low heart rate, excessive slow wave activity in the brain, problems in paying attention, and abnormalities in the frontal area of the cerebral cortex.

The rival theory of environmental influence, however, has been given considerable weight by psychologist Adrian Raine, who points out that criminals are more likely than the rest of us to have criminal parents; to have poor marital relationships; to batter their wives and abuse their children; and to have been subjected to unpredictable parental punishment.[308] Raine and his colleagues surveyed 4,660 boys in California, all born between 1959 and 1961, and found a clear connection between violent criminality and early experience. They looked at birth complications such as breech and forceps deliveries (which can damage the brain), and parental neglect during the first year of life, and found that 3.9 percent who were exposed to *both* such traumas committed twenty-two percent of the group's serious crimes. Raine concludes that better care for pregnant women and newborn babies may be more effective than outlawing guns or having more policemen. "Biology," he says, "doesn't have to be destiny."[309]

Just so. Our behavior is most often the result of a complex interaction between nature and nurture, between inheritance and experience. We are astonishingly plastic in such matters, drifting in the direction determined by natural selection, becoming predictable where resources and access to them are involved. Theft, burglary, assault, homicide and even war can often be described in sociobiological terms and given an evolutionary significance. I have tried through most of this inquiry into evil to show that such things often have clear biological roots and can contribute, in the final analysis, to reproductive success. These perhaps can be defined as "weak evil" in the sense that they are, even if only weakly, adaptive. But we also have a tendency to behave, from time to time, in very unpredictable ways that are clearly maladaptive. Arson, pedophilia, and the pathic peculiarities of serial murder are beyond logical or biological comprehension. They defy every reasonable attempt to make sense of them, though we are bound, as scientists, to try to understand uncomfortable facts such as the overwhelming tendency of serial killers to be white men. Attempts to do so are bedeviled, of course, by the equally awkward fact that those who do such things are frequently as bemused by their behavior as the rest of us.

I have no intention or inclination to dwell on the sadistic horrors of much serial killing. These have been well documented in Jack Apsche's chilling probe into the mind of "Bishop" Gary Heidnik who abducted, raped, tortured, impregnated, killed, carved, cooked and ate his victims in Philadelphia a decade ago.[12] A psychiatrist's report on him in 1987 noted that: "with continued psychotherapy, Mr. Heidnik's prognosis is good." But at that very moment, he had three naked, abused, malnourished women chained to the plumbing in his basement, and the remains of another in his freezer. There are painstaking analyses of the charismatic Ted Bundy by Ann Rule;[329] of lonely Dennis Nilsen by Brian Masters;[266] and of the insatiable Jeffrey Dahmer by Anne Schwartz.[340] But however you look at it, murderous misogyny and homosexual necrophilia are unattractive and unproductive. There is little in any of these studies to help us understand what such excesses are really about. Although Nilsen's reply to a question about the motive for his fifteen killings may come as close as we are likely to get. He said: "Well, enjoying it is as good a reason as any."[266]

Many of these killers appear to have no inner being. Speaking of John Wayne Gacy shortly before he was executed, Alec Wilkinson said: "I often had the feeling that he was like an actor who has created a role and polished it so carefully that he had become the role and the role had become him. What personality he may have had collapsed long ago and has been replaced by a catalogue of gestures and attitudes and portrayals of sanity."[416] That sounds absolutely right. A description of Robert Thompson and Eric Smith twenty years from now, if they live that long and learn how to cloak and dissemble, how to put a layer of human makeup over the "Other Look." In all their cases, it might be more productive to look at what they are *not,* instead of what they are or have become.

I don't believe that anyone is born with a compulsion to kill. And while I am impressed by the wealth of evidence that links poverty, brutality and social deprivation to crimes of violence, I am far from certain that anything specific goes into the *making* of a murderer. For every one who does end up killing, there are hundreds from the same gene pool under the same dire circumstances who don't become psychopaths. I am beginning to feel sympathetic toward those who have come to reconsider the idea that evil, "strong evil" of the kind we may have been dismissing as mental illness, might exist after all and, if it does, I suspect that I may have seen its face.

Joel Norris suggests that serial killers, at the moment of their consummation, have "blinding revelations of truth." Dennis Nilsen said: "I have no other thrill or happiness." It is clear that murder represents some sort of peak experience in the otherwise bleak lives of these killers. Novelist Joyce Carol Oates sees in it a profile of "a bizarre species of romantic love, the project of a ghoulish Don Juan whom no quantity of bodies, or body parts, will satisfy."[290]

Perhaps. But I am still haunted by the "look," which seems not to be a phenomenon peculiar to the second half of the twentieth century. It stares back at me from the marble busts of Caligula and Nero, but not from those of Julius Caesar or Ptolemy. I see it in the paintings of Cesare Borgia and Torquemada, but not in the likenesses of Dante or Petrarch. And it leaps out at me from photographs of Rasputin and Idi Amin, but I can find no trace of it in the many images of Winston Churchill or John F. Kennedy, whatever their other faults. And it seems to me that each of those who has the Other Look has also personally succeeded in something unspeakable, something so counter to natural history that it sets them irrevocably apart, cut loose from the pleasure and the pain of being truly human.

This is how I believe it came to be.

Three billion years ago, we slipped across an organizational divide, resisting entropy for long enough to pluck order out of inorganic chaos, becoming recognizably alive. We became part of a process which may be unique to this planet, but follows universal laws, seeking fleeting equilibrium, finding balance in patterns with certain harmonic properties, avoiding those that were not "just right" for the ecology. This is what is meant by "good." It is naturally beautiful because it fits so well, and that "fitness" survives because it is selected for in competition against those innovations that don't work as well, and are inevitably ugly.

Ugliness, that which is maladept, gets to be that way because it has lost its place, lost its balance, or lost the diversity necessary to provide appropriate answers to the ever-changing challenges of evolution. The prime mover in this long battle for selection has been the genes, the replicators, who give their vehicles the instructions they need for reproductive success. For 99.99 percent of our life history, they have been in charge: our founding fathers, running a very successful business by remote control, setting up systems of such intricacy and delicacy that they literally take one's breath away.

But 3 million years ago, something different happened and we crossed another great divide. This process threw up a permutation that looked weak. It didn't have any of the usual assets of strength, speed or subtlety, but it turned out to be a winner because it could compensate for its lack of physical prowess by an extraordinary mental agility. And that changed all the rules. These newcomers, our immediate ancestors, continued to benefit from genetic inheritance, starting each individual race a good way down the track, piling up advantages until they put sufficient distance between themselves and the rest of the pack to become the dominant life forms, top species, *Homo sapiens.*

Our specific history, though it occupies but 0.01 percent of life's whole, has been remarkable. There is no truly objective basis on which to elevate one species above another, but it has become obvious that ours is qualitatively different in at least one important way. We have the power to defy the genes. We have questioned their authority, rebelled against chemical control and, even before we knew who or what they were, set in train a movement which represents a real alternative to their tyranny. We have invented cultural evolution which, compared to the biological process, happens at the speed of light.

It takes a long time to change a leg into a wing or a flipper, but we can change our minds and our destinies in a second. The advantages of this quantum leap are obvious and enviable—all the blessings of art and music, poetry and philosophy—but there has been, as there always seems to have been, a price to pay. We have lost touch with our roots and, like Antaeus—whom Hercules killed by holding aloft, out of contact with the strength-giving Earth—our natural powers have begun to ebb. We may be in charge now, but we have lost control of the old checks and balances, the essential inhibiters and disinhibiters, which make for easy equilibrium. And we seem to be in danger of losing the Goldilocks touch altogether.

The break with organic evolution and genetic control has all happened so fast that there hasn't been time to put appropriate cultural controls in their place. And the result is a touch of anarchy, an outbreak of actions which can best be described as cosmic heresy, natural blasphemy and ecological sacrilege. A growing catalog of crimes against nature, not the least of which is the assault on our own psychic integrity, turning some among us into something else, something now barely recognizable in human terms. Some thing that peers balefully back at us with its Nosferatu face.

There may be a simple psychological explanation for this Other. Killing in cold blood is an act so terrible that the only refuge from it may be in denial so complete that personality slips away altogether. This may be the only way in which a killer can survive. As in Asmat, survival requires death. Only this sacrifice seems not to be one in which the man gives way gracefully to the child, but one in which the child dies to allow the man to go on, albeit in some almost unrecognizably monstrous form. That could even be the gene's last desperate stand against our revolution.

We live in extraordinary and fascinating times, poised on the brink of something either too wonderful to imagine or too terrible to contemplate. Probably, given what we know of life, a bit of both. But the mix remains vital and possibly still within our powers to control. What worries me most is that we have an alarming tendency to tilt the balance in the wrong, the ugly way. How else can one explain the fact that mass murderers have now become cult figures celebrated in an unsavory market of killer kitsch? Of trading cards, posters and T-shirts that promote such adulation of serial killers that Ted Bundy even found one groupie credulous enough to marry him. Man-eating Jeffrey Dahmer, until his own murder in a Wisconsin prison in late 1994, was showered with fan mail and thousands of dollars in gifts. Richard Ramirez, the "Night Stalker" who savagely murdered thirteen people in California, even now has a devoted following of women who write to him and even visit him in San Quentin. And there are those who have begun to turn to Charles Manson for advice.

Some of the attention is inevitable. We have always enjoyed flirting with danger, scaring ourselves with the possibility of pacts with the devil and dances with werewolves. In a way, such perversity is even healthy. It helps to remind us of who we are, and of the risks of becoming the Other. But by fanning the flames of Hell we are giving life to a truly errant meme and flirting with real problems. I don't know if Robert Thompson saw Chucky's video, and I doubt that Eric Smith ever heard of James Bulger, but such direct influence may no longer be necessary. I suspect that there is a biological version of Internet to which we all subscribe, and there may now already be enough bad news in the air to infect anyone, anywhere, at any time.

PART THREE

EVIL NATURE

SIN IS WHATEVER OBSCURES THE SOUL.

André Gide

No rain that falls in the Great Rift Valley of East Africa ever reaches the sea. It collects instead in a necklace of lakes on the flat bottom of a narrow trough with sides so steep that the geological name for such a feature is graben—a grave. There are bones and fossils here enough to satisfy any paleontological ghoul, because this was not just where our ancestors died or were buried, but where our whole lineage was born. It is as much cradle as grave, and has a very pleasant and familiar feeling to it.

It is home now mainly to pastoral people—Maasai, Samburu, Suk and Turkana amongst them—who have driven a Nilo-Hamitic wedge down into the heartland of agricultural Bantu territory in the cool, fertile highlands on either side of the valley. Herders and farmers here eye each other with inevitable suspicion, but their languages and lifestyles have begun to converge, and there is one thing anyway on which they are all in full agreement. Neither has much time for a marginal group of hunter-gatherers both know as *Wa Ndorobo.*

If you run the sounds together, you get Wandorobo, a name which, from the moment I first heard it, conjured up images of nomadic groups of pre-invasion aboriginal people still armed with bows and arrows, sleeping in caves and simple shelters, collecting honey and roaming the escarpments at will. Natural man, unchanged for 100,000 years. The truth, I discovered when I went to live in the Rift, is not quite so romantic. They do indeed live a little like that, but have long since lost their ethnic purity, mixing with local farmers and herders all along the valley, picking up their languages, swapping meat and honey for trade goods, rapidly becoming more reliant on the twentieth century and its dubious services. But, if the Dorobo no longer live as real "people of the bush," their symbolic role is constantly being reimagined. If they did not exist, they would have had to be invented.

To herders and farmers, the humanity of the Dorobo, who neither farm nor herd, is questionable. They live beyond the constraints of "normal" society and property, and are therefore difficult to classify or control. "The Dorobo are like animals," they say, "yet they have power over nature; they live in the realm of witches and wild spirits, but also in the presence of the ancestors; they are profligate, amoral fools, but are also the possessors of secret knowledge and are able to mediate with powers beyond our world."[227]

For almost everyone in Tanzania, Kenya and Uganda, the Dorobo provide an essential foil, a mirror in the forest, a necessary image of the Other. Every society seems to need an Outsider, against which the qualities of Insiders are contrasted and confirmed. To the Maasai, who attach great virtue to the accumulation of cattle, the Dorobo are those absolutely mad creatures who even "eat the heifer." For the Kikuyu, who measure their wealth in land, Dorobo are worthless people who "live in the forest like baboons." And among the Kipsigis, where a man is known by his manners, Dorobo are dismissed as "those who do not know how to behave."[326]

All of which is very convenient. If we, in our own eyes, are the advanced members of a modern and progressive society, then we must have progressed with respect to something else, such as the "primitive" Dorobo. If our culture is riddled with invidious class distinctions, then at least we can find solace in the thought that the Dorobo have no society to speak of at all. If we are immersed in a cycle of violence and random killing, it is comforting to know that, like the "gentle" Dorobo, we were once much nicer to one another. All that we are, whatever that might be, the Dorobo of the world are not.

It is fascinating, in this respect, to discover that the infamous Ik—anthropologist Colin Turnbull's *Mountain People,* whom he found to be loveless, savage, selfish, destructive and inhumane: "as unfriendly, uncharitable, inhospitable and generally mean as any people can be"—are a clan of Dorobo.[390] People with whom he went to work reluctantly, when two other preferred projects fell through, and whom he found in a state of unusual impoverishment and dissolution as a result of being driven from their ancestral hunting grounds by the establishment of a new national park. His reaction to the Ik was tempered, he admits, by disappointment and by a lack of enthusiasm for being there at all. He had occasion, after his first few difficult months amongst the Ik, to write a chapter about them in another book. And that differed so significantly from his later assessment that he said: "It makes me question the validity of much fieldwork, including my own." But in a real sense, disillusion was probably inevitable, because the Ik and all other Dorobo have for so long been marginalized and described in such totally negative terms that they have become what most people expect them to be. Other-wise.

And if there isn't a convenient scapegoat like the Dorobo nearby, societies are adept at manufacturing one of their own on the inside. There are some easy targets.

The trade of the blacksmith in many, if not most, tribal societies has become strongly stigmatized. Messing about with metals, forging iron from iron ore, giving birth to new matter before its proper time, has always been seen as tempting fate. Something best done by a special caste of men, hereditary "black" smiths whose constant contact with the occult was considered so dangerous that they had to be held at arm's length. And to reinforce this distancing, the Kapsiki of North Cameroon have created the convenient myth that blacksmiths "smell bad."[393] From there, it is just two short steps through "bad smelling things must be bad," to the very widespread practice of the "smelling out" of dangerous individuals—for the finding of witches.

Ever since there were humans, there have been human desires.

These differ from culture to culture, but the one fact common to all people everywhere is that nobody's desires are ever completely satisfied. We always want more. More food, more land, more women, more children, more power, more respect, more money, more freedom—whatever. And as each desire is temporarily sated, the need transfers automatically to another kind of hunger. In most other species, needs are constrained by ecological realities and, though competition persists, communities settle into more or less stable sets of relationships. Humans changed all that by turning natural needs into more disruptive "wants" and set in train our most universal characteristic, a tendency to disturb the ecology by trying to control it to our own advantage.

Our first attempts were probably directed at increasing productivity, and took the form of projecting our own internal desires onto the external world. Something we seem to have begun to do very early, perhaps as soon as we gave up scavenging and learned to hunt.

Hunting is a risky business, a hit-or-miss affair, and anything that seemed to lessen the uncertainty would have been very welcome and quickly adopted. An American anthropologist with the magnificent name of Omar Khayyam Moore once suggested how this could have begun.[276] He pointed to a custom which may still be in use among Native Americans who hunt the caribou in Labrador. When uncertain which way to take the hunt, they consult the hunted. They hold the shoulder bone of a caribou over hot coals and then interpret the cracks and spots produced by heat as a map, going where it seems to indicate. The marks are largely random, and are likely to lead to caribou only as a matter of chance, but over a period of time those using the technique and making such random forays tend to make better and more balanced use of their land. And in the long run, they become more successful hunters than those whose natural inclination to go back to where they had last hunted successfully very often leads to local overhunting.

There is a tendency to dismiss such practices as "superstition," with the condescending observation that they are based on false premises. Perhaps. But as Moore points out: "Some practices which have been classified as magic, may well be directly efficacious for attaining the ends envisaged by their practitioners." Sometimes the magic works.

Success breeds success, and such customs quickly spread to other needs in search of fulfilment. Before long there are rituals, accompanied by appropriate verbal formulae, for catching fish, making rain, taming the wind, finding love or water, and ensuring fertility. Some of them based, by accident or design, on cues which have a natural relevance, and soon we are knee-deep in spells and all the evocative techniques of divination designed to make sense of the accidents of life. I must admit, in passing, to a personal preference for apantomancy—which finds significance in chance meetings with animals; and for phyllorhodomancy—an analysis of the variety of sounds that can be made by slapping rose petals against the hand. But the point is that, very early in our history, we became adept at, and addicted to, productive magic of the kind that is dedicated to a particular end.[126]

These acts of magical control must have begun as personal ones, performed by individuals or small groups for their own private ends. But those that could be seen to work will very quickly have won social approval and become rituals performed for the community as a whole, eventually by specialist magicians with recognized talents for doing so. This state of affairs is one which existed until very recently among living hunter-gatherers such as the Australian Aborigines and the Kalahari Bushmen. But with the transition to a more settled, agricultural way of life, there came major changes.

At first, these were ones only of kind, reflecting a new way of life in which desires turned more to the need for securing the harvest, guarding property, finding safety outside the home, averting misfortune, and making trading profits. This is protective rather than productive magic, the direct result of a more defensive lifestyle; and initially it was practiced in the same way, with social approval and by specialists in sorcery who soon came to be a powerful force for social control.

But then, as new opportunities for wealth and power, greed and envy, were created, they also spawned a totally new kind of magical enterprise. Destructive magic. Magic designed and employed to bring down storms big enough to destroy a rival's crops or lay waste to his property, to produce sickness and even death. This has never had social approval anywhere and, in many societies, never even been practiced at all. It was always classed as morally bad and frequently diagnosed only on its supposed effects. Those who did practice it never admitted to doing so, because this was not just sorcery, but witchcraft.

The distinction is one first drawn by pioneer anthropologist Edward Evans-Pritchard in his classic study of the Azande in the Sudan. He said: "Azande believe that some people are witches and can injure them in virtue of an inherent quality. A witch performs no rite, utters no spell, and possesses no medicines. An act of witchcraft is a psychic act. They believe also that sorcerers may do them ill by performing magic rites with bad medicines."[123] The situation is less clearly defined elsewhere, with witchcraft taking on a whole range of functions from providing love charms to holding Satanic black masses, but the distinction remains a useful one. Sorcery has social approval; and witchcraft doesn't. Sorcery involves technical skills whose use can be learned by anyone; but witchcraft is something more personal, practiced by an adept born with or given the special talent to do so. Sorcery can be seen; while witchcraft, almost by definition, cannot.[252]

And yet, a belief in witchcraft is so widespread and its character so similar from Africa to America, from the South Seas to Asia, that it must be related to something real in human experience. The answer seems to be that it provides a native theory of failure, misfortune and death. It allows us to believe that the blame for at least some of our suffering rests upon a peculiar power which can be exercised by certain people in our midst. By witches, who are usually thought to be adult, often female, perhaps with visible marks; to be stingy, reserved and quarrelsome; to work from envy, malice and spite; to operate in secret, usually at night; and to turn their evil talents against their own neighbors and close kin, never against strangers. They are, in short, our own ugly relatives, the ones we most love to hate. That part of ourselves that makes us most uncomfortable.

Witches are also seen as not entirely human. Their evil power is something that comes from possession by spirits or devils, by "snakes" in the abdomen, which give them nonhuman power. And by their very nature, witches customarily reverse all normal standards. They indulge in "unnatural" practices such as incest and bestiality; they dig up the dead and eat their own children; they go naked instead of clothed; they walk backward and ride facing the tail; and they recite sacred texts in reverse. They are, once again in brief summary, a perverted picture of ourselves, everything we become afraid to see when we look into a dark mirror.[267]

There may be individuals who have the courage to accept personal responsibility for such shortcomings and become practicing witches on that account. But most of us are less brave and far more ready to grasp at beliefs which not only absolve us of blame, but provide remedial action in an approved social form, which can even result in having someone we truly dislike publicly burned at the stake, or even crucified. Someone else who dies for our sins.

In most known cases, the witch and his or her accusers are closely related. They are individuals who ought to like each other, but do not. They are bound genetically to be nice to one another, but feel temperamentally incapable of doing so—and that produces a very painful and dangerous tension. An accusation of witchcraft resolves that dilemma and provides a convenient excuse for rupture and release at the personal level. It also very often is cathartic for the whole community, providing a public enemy that can be ritually dispatched, allowing everyone else to breathe more freely, especially if the witch can be seen to embody all those characteristics which that society finds most distasteful.[242]

In tribal society, particularly in Africa, witchcraft is a handy explanation of evil. It gives people the chance to account for their failures and frustrations. The source of the evil is generally supposed to be nonhuman, but as it operates through human beings, it is possible for other humans to bring it under control. Something can be done. Moral order can be restored. The purest form of such belief is to be found amongst the Lobedu in South Africa, who see the motive force behind witchcraft as an intrinsically evil influence in the universe. Something quite separate from all other spirits or supernatural forces. Something so mysterious that it transcends nature; but also something which can manifest itself only through a human being, where it makes use of the worst impulses of the human heart.[232]

This wonderfully intuitive understanding provides the Lobedu and many others with an elegant, very human, way of labeling and dealing with their overt and covert desires. It recognizes evil and acknowledges its place in our lives, but all the time emphasizes its personal nature by concentrating on the witch and the bewitched. Such allegations and actions are seldom allowed to become general or to escalate into collective patterns of infection or persecution. That would never do.

In industrial society, we handle such things very much more awkwardly. Largely because of our monotheistic tradition.

There were always tribal gods. Lots of them. Gods of fire, earth and water; gods of life and death; gods of war and gods of the crops. There were good gods and bad gods, kind gods and mean ones. Ones you could count on, and others you would go a long way to avoid. They were usually invisible, but nevertheless very human in their desires, easily angered and propitiated only with difficulty.[372] Sometimes there was a god of gods, potent enough to stand above the others, but it wasn't until Yahweh, the Hebrew Lord of the Universe, that any deity was seen to be absolutely omnipotent. Without parallel, beyond compare, and apparently without opposition. The idea that anything could take place without his participation was theologically intolerable. Revisionism however set in during the second century B.C. with the first mention of *Satan,* meaning "adversary," as a proper noun and as an evil force in direct opposition to God.[192] But from that moment on, and principally in apocalyptic and apocryphal texts, Satan—sometimes known as Belial or Mastema—becomes a major player, at war with God at the head of an army of lesser demons.[230]

That moment also marks the beginning of what is seen as a prodigious struggle between Satan and the new religion of Christianity. The kingdom of Jesus against the kingdom of the devil. And, as usual in such conflicts, anyone who was not for Christ was against him; which put Jews, Muslims, and all others perceived as "pagan," among the ranks of the devil's supporters. It is all part of a very familiar pattern.

We need, it seems, to give evil a face. Evil is never abstract. It is something done to an individual by someone or something, and we need very badly to put a face or a name to that force in order to bring it under some semblance of control. Imagine a mother with a baby in her arms. She smiles, the baby smiles. She laughs, and the baby laughs with glee. Then suddenly, without warning, and for no apparent reason, the baby's head explodes. Horrible, but what does it mean? Was it her fault? Are they bewitched? There are no easy answers—and that, in itself, is terrifying.

Now try Dostoyevsky's version in which Vanya Karamazov is discussing evil with his younger brother Alyosha: "Imagine a trembling mother with her baby in her arms, a circle of invading Turks around her. They've planned a diversion; they pet the baby, laugh to make it laugh. They succeed, the baby laughs. At that moment a Turk points a pistol four inches from the baby's face. The baby laughs with glee, holds out its little hands to the pistol, and he pulls the trigger in the baby's face and blows out its brains."[111]

Still horrible, and perhaps made even more chilling by the artistic flourish of the "diversion." This is unquestionably evil, but the fact that it has a Turkish face and a focus, someone to blame and hate, perhaps even to hunt down in vengeance, gives the horror a human, manageable dimension. It is easier to do that than to accept that the world itself is an evil place from which there is no escape but death. So we pass responsibility on to convenient scapegoats. We conjure up demons and evil spirits to set against our "good" gods, and we invent traitors in our midst to account for the fact that some of us seem not to be on the side of the angels.

British historian Norman Cohn has taken a long hard look at the whole story of witchcraft in Europe; of beliefs which inspired the great witch-hunts of the sixteenth and seventeenth centuries, in which thousands of apparently innocent people were burned alive or tortured to death. And he comes to the conclusion that it was just a story, that witchcraft in Europe has little or no historical basis whatsoever.[75]

Witch-hunting, on the other hand, was very real. It emerged from a state of confusion in the Christian Church at the end of the first millennium, when peasant and military uprisings added to general feelings of dissatisfaction. Both Church and state moved to quell the disturbances and found that the easiest way to do so was to shift responsibility for the crisis to imaginary demons and to those in which such evil spirits were believed to take human form. They gave evil a human face, and anger and frustration an explicit focus. They turned old women, who were expendable anyway, and often seen to be troublesome and eccentric, into perfect low-cost scapegoats. And with forgeries and lies they whipped up a religious frenzy that lasted for centuries and very efficiently drew the poor away from confrontation with the establishment, and away from growing demands for the redistribution of wealth and the leveling of rank.[176]

What began as cynical manipulation grew into a hunt that spread from witches to all heretics, anyone who questioned the *status quo.* "Those who believed that there were devil-worshiping societies in the mountains, soon discovered that there were devil-worshiping individuals in the plains."[385] All were tarred with the same brush, charged with the same crimes and executed with the same brutality. When ordinary rules of evidence and all civil rights are suspended, when unremitting torture is used to obtain confession to impossible acts, when the only trial that takes place is a trial by ordeal, no one is safe. The authorities could and did dispose of anyone they wanted to get rid of, such as Joan of Arc, and they did so in the name of God. By creating a powerful social myth, they turned fantasy into history.

Cohn points out that there is much in the nature of the fantasy which looks like the eruption of obsessive fears and terrifying desires. He calls the demons "endopsychic" and suggests that the imaginary society of witches represents a part of our own innermost selves. "No great psychological sophistication is required to see that the monotonous, rigidly stereotyped tales of totally, indiscriminately promiscuous orgies do not refer to real happenings but reflect repressed desires or, if one prefers, feared temptations . . . When a real or imaginary outgroup is accused of holding orgies of this kind, it is certainly the recipient of unconscious projections."[75]

That sounds right. Christianity has generally tended to exalt spiritual values at the expense of natural desires, and repressions do have a way of bouncing back in monstrously distorted forms. The witch-hunts began at a time when there were already strong anticlerical feelings and an unconscious revolt against religion itself. And it would be surprising if these had not produced the sort of tension that could find an outlet only in sympathy for the devil, and in fantasies about erotic debauches with him. There is, however, a limit to such Freudian insights. It is unlikely that even the most repressed felt compelled to gather in covens at sabbats where babies were slaughtered, their blood drunk and their flesh devoured. There have probably always been individual psychopaths who will do such things, but the notion of cannibalistic, orgiastic, devil-worshiping sects was as fantastic in the Middle Ages as it still is now. It is one of those fantasies to which we seem prone, but it is worth asking where the fascination for such things lies.

Snow White only just escapes in time from the wicked queen who eats the lungs and liver of little girls. Hansel and Gretel are captured by a repulsive hag who kills and eats children. The Titan Cronus tries to swallow his own children until the youngest, Zeus, overthrows him. But the evil seed flourishes as Zeus's own son, Tantalus, cooks his son as a meal for the gods. Cannibalistic infanticide seems so much on our minds that psychoanalytic theory has had to wonder whether we do in fact harbor such unconscious impulses, and whether children could be subliminally aware of them.[107] This sounds like desperate theorizing around a common tendency among fond relatives and strangers to leer into cradles, and for besotted mothers to nibble their babies while saying silly things like: "Oooh, I could eat you all up!"

There is little to worry about in that, but there is a genuine generational disparity based on the genetic conflict of interest between parents and offspring—and that is worth taking seriously as a source of tension that may well have contributed to the stereotype of witchlike behavior.

There may never have been any real witches in Europe. The evidence for them anywhere at all is very slim. But what the witch myth recognizes is the real opposition inherent in moral value systems. There is an inevitable conflict between right and wrong, proper and improper, good and evil behavior. We feel this in ourselves, we get tugged each way and end up more than a little confused. But by putting someone else, a scapegoat of any kind, on the wrong side of the local moral line, we do ourselves a favor. We put distance between us and them, between insiders and outsiders, between ourselves and the Other.

It is no accident that those accused of witchcraft should so often be close relatives, traitors within the gates. Their closeness emphasizes our own involvement and our peril. But it also turns sin into an unforgivable crime, that of betrayal. And this cries out for the maximum penalty—which also happens to be the only one which will satisfy the genes and ensure that the traitor, the nonreciprocator, never gets to pass on his or her particular genetic aberration. There is no harsher judge than the gene whose fitness is compromised, and no easier way of spreading delusion than by appealing to some imagined genetic slight.

The last witches in modern society may have been put to death in Massachusetts in 1692 and in Switzerland in 1782, but we still have witch-hunts. We go on persecuting those who seem to be a threat to our party, our Church, our race, our sex, or even something as tenuous as "the American way of life."

In the United States in 1950 the threat of Communist infiltration seemed real. The movement existed. Its aims, tactics and direction were mysterious, and fear of it was considerable. Whatever the reality of the threat, that fear alone was what made Joseph McCarthy possible. He began by making airy warnings about "Them—the enemies of the people"; and once he was immune from libel on the floor of the Senate, claimed to have the names of 205 "card-carrying Communists" in government. His cynical lies and blatant opportunism, his pretense of patriotism, seem obvious now, but his position as a senator made a difference at the time. People took him seriously. His pursuit of "pink shadows" and "red slime" through the corridors of power created fear and suspicion throughout the country. With his own spy system, helped by people who should have known better, and by skillful use of the repeated big lie, he destroyed lives, paralyzed the will of two Presidents, and interfered with the nation's foreign policy to an astonishing extent.

Not one of his accusations was ever substantiated, but for several years this unscrupulous man without party, program or organization succeeded in distorting truth and decency just as the Inquisition had done four centuries before, preying on people's fear simply of being accused—this time of treason, the twentieth-century equivalent of heresy. The spell was eventually broken and McCarthy censured and stopped. But something very like the witch-hunt that he organized has appeared again, just forty years later, in a campaign against President Clinton and his wife, in which the demonology is identical: "Do-gooders, the Democratic leaders, the liberal intelligentsia, and a decadent Eastern press."[124]

We need enemies. Without them, we would only have ourselves to blame. With them, we have a convenient receptacle for all the qualities that we find it hard to tolerate in ourselves. It is easy to break the rules and steal a march on drivers in other lanes on the freeway, but so much more satisfying to hate the other guy when he squeezes in ahead of you. Self-righteousness feels good. We are right; they are wrong. We are innocent; they are guilty. We tell the truth; they lie. We have a defense department; theirs is called a ministry of war. We make enemies of them, not because we are inherently aggressive; but because striking out at them brings us closer together. Being close feels good, so we do what seems necessary to be good. We make more enemies. We manufacture evil.

"We must beware," said Carl Gustav Jung, "of thinking of good and evil as absolute opposites." They are not, anyway, always self-evident. Much, as I have tried to show, depends on the context. On what is good for the ecology. The greater good quite often means doing what would otherwise be considered as evil. It was Jung who identified the darker side, the part of us we have no wish to be, as the *Shadow.* It is sometimes seen simply as a hold-all for disposal of inconvenient parts of the psyche; poet Robert Bly calls it the "long bag we drag behind us."[43] But Jung saw it as something more independent; that part of our consciousness which coalesces into a separate, yet well-defined, inferior personality which is usually concealed from exposure and discovery. He saw it as an essential part of our complete being, so vital that without it we would disintegrate and die. And much of Jungian therapy continues to concentrate on ways of recognizing the Shadow, of bringing it to conscious attention and using its ideas, values and judgments to complement our own. "Who knows what evil lurks in the hearts of men?" The answer is easy, of course. The Shadow knows.[218]

Jung was convinced that the Shadow was more than a symbol, that it really exists: "It is as evil as we are positive . . . the more desperately we try to be good and wonderful and perfect, the more the Shadow develops a definite will to be black and evil and destructive . . . The fact is that if one tries beyond one's capacity to be perfect, the Shadow descends to hell and becomes the devil. For it is just as sinful from the standpoint of nature and of truth to be above oneself as to be below oneself."[219]

Just as sinful? Yes. With this, we come back to Aristotle's idea of "just enough," not too much or too little. Evil is not simply an absence of good, it is an absence of balance with good, and becomes manifest by such disequilibrium in a human being. The only real danger perhaps is already in us, not lurking in some infernal pit. There is little merit in searching for evil in witches, devils, demons or other scapegoats. We are it. In the words of cartoonist Walt Kelly's *Pogo:* "We have met the enemy, and he is us."

But that doesn't mean we can't fight it.

SIX

THE MARK OF CAIN: THE IDENTITY OF EVIL

I began this exploration of evil with the comment that things in nature have "a glorious and unsettling tendency to be beautiful." They do, and they truly are glorious, but now I must explain the unsettling part.

I find the beauty sometimes disturbing because, as a biologist, I know that nature is basically relentless and unfeeling, with little in it to match the sympathy of a child who stops and takes the time to turn a beetle back over on its feet. It's a jungle out there, a war very often of all against all, in which ten percent of all known species are parasites, whose job it is to harass, weaken and disfigure the others. Part of what writer Annie Dillard calls life histories "in some hellish hagiography."[109] She is right, such things are not well enough known. The romantic, carefully censored world of televised nature conceals the brutal truth, which is that most of the creatures on this planet live in constant and justified fear of the rest, or pay their way as slowly dying hosts to unthinkable lodgers.

"The universe that suckled us," rails Dillard, "is a monster that does not care if we live or die." And that too is true. Nature has little, in a moral sense, to commend it. The Victorian scientist and humanist Thomas Henry Huxley, who did battle with Bishop Wilberforce on Darwin's behalf, decided later that: "Brought before the tribunal of ethics, the cosmos might well stand condemned."[208] He delivered this *fin de siècle* verdict in a lecture at Oxford University in 1893, two years before his death. The Victorian age too was drawing to a close and with it went some of the imperial certainty about right and wrong, and about man's proper place in nature. The Church was in disarray, everyone was looking for new sources of moral authority, and Huxley was determined to make it clear before he left that such lessons were unlikely to be learned from nature. On the contrary, he insisted: "Let us understand, once for all that the ethical progress of society depends, not on imitating the cosmic process, still less in running away from it, but in combating it." This redoubtable man they called "Darwin's Bulldog," not content with creating a last and haunting image of human isolation, went out with a ringing declaration of war.

The enemy in this case was not nature itself, but that part of it that Huxley saw as morally indifferent. As an antidote, he urged adoption of a new ethical strategy: "In place of ruthless self-assertion it demands restraint; in place of thrusting aside, or treading down, all competitors, it requires that the individual shall not merely respect, but shall help his fellows; its influence is directed, not so much to the survival of the fittest, as to the fitting of as many as possible to survive."[208]

There is evident in this call to arms a good bit of early Victorian bravura, and a grasp of genetic priority more profound than that possessed by many scientists today, a full century later, even with overwhelming recent revelations about rampant self-interest in parts of nature that were once seen as relatively benign.

We realize now that the details of family life can be so murderous that even a seasoned naturalist such as Sarah Hrdy, who came to know all the langurs in her study area as individuals, found herself sometimes weeping as she worked.[196] Robert Trivers has shown that parent-offspring conflict is so widespread that it should be regarded as the expected norm in most family life.[387] Offspring are bound to try for more than their fair share of resources, and parents are equally committed to a compromise which might mean starving one or more of their young to death. And even when sufficient food is available, all mammals face a period of acute strife over when weaning should take place.[4]

Courtship in animals, which once seemed to be simply a matter of stately dances and colorful displays, can now be seen to be a web of intrigue full of deception, coercion, desertion, rape, and sometimes lethal violence. Largely because of conflicts arising from the disparity of investment between the sexes. Even a tank of tropical fish, once a popular refuge for those looking for relief from human stress, has been revealed to be a scene of never-ending underwater sexual harassment.[88] And bird songs, those "profuse strains of unpremeditated art," stand revealed now as fragments of an ongoing domestic argument based on male suspicion and female protestations of fidelity.[156] Part of a universal arms race between the sexes that makes promiscuity a useful hedge against infertility in a partner, and cuckoldry commonplace. In one study, half the mates of sterilized male blackbirds succeeded nevertheless in producing fertilized eggs.[324]

Among mallard ducks, aggravated sexual assault is such a normal part of reproductive behavior that an unguarded female may be gang-raped so persistently that she drowns.[25] Man-eating male blue sharks seem to be unable to mate until they have made injurious attacks on their own females.[307] Rape is an established pattern of behavior in many insects, frogs and turtles; and can even be homosexual in some parasitic worms.[1] Homosexual behavior generally is rampant in the salamander,[293] and very common in a wide variety of other vertebrates, for some of whom it may even be the practice of choice.[409]

Cannibalism has now been recorded in just about every animal that isn't strictly vegetarian. Female spiders do it routinely as a way of giving their young a nutritious start in life, and male wasps on the make are courting death by approaching a female who is already mated.[56] One in every twelve ground squirrels in a colony is killed and eaten by its own male kin; while female squirrels seem to be content with just killing, but not eating, all young in the nests of their competitors.[353] Male fish entrusted with guarding the family eggs may, if they are hungry enough, simply eat them; and anyone who has kept an aquarium knows how important it can be to separate young fish from their parents before similar temptation prevails.[112] Large Canadian perch called "walleye," because of a sinister opaque look they cast even over members of their own species, have been found with smaller walleyes in their stomachs, which in turn had eaten still smaller ones, and so on for four unrepentant generations of cannibals within cannibals.[84]

Fighting between rival males ought to be ritualized and largely risk-free, but the sad truth is that it can also quite often be lethal. As many as one in ten red deer stags dies each season from such injuries.[73] Edward Wilson estimates that roughly once in every thousand hours of field study a human researcher is likely to see one of the animals in his area kill another of its own kind—and contrasts this with the chances of observing one human kill another in, say, downtown Washington or Houston, currently the most murderous of major American cities. It seems that to see one out of ten chosen Houstonians die violently, it would be necessary to keep that group under constant surveillance for over 300 years. "I suspect," says Wilson, "that if hamadryas baboons had nuclear weapons, they would destroy the world in a week."[426]

The conclusion to be drawn from this unhappy catalog of unpleasant things is obvious. Huxley was right. Nature is morally bankrupt and stands condemned. Worse than that, says the professor of biology at the State University of New York, in light of recent discoveries, nature is clearly and grossly immoral. "No one of Huxley's generation could have imagined the current concept of natural selection, which can honestly be described as a process for maximizing short-sighted selfishness. I would concede that moral indifference might aptly characterize the physical universe. But for the biological world, a stronger term is needed." The only one that will do, it seems, is evil.[420]

George Williams is a Lincolnesque figure, quiet and unassuming, but with considerable presence, the godfather of evolutionary biology. In 1966 he wrote a book called *Adaptation and Natural Selection,* which explained for the first time how compulsive self-interest, despite all its immoral connotations, can be a very successful strategy.[418] He defined natural selection in Darwinian terms, looked at it warts and all, and concluded: "Mother Nature is a wicked old witch." He also decided that: "Natural selection really is as bad as it seems and that, as Huxley maintained, it should be neither run from nor emulated, but rather combatted."[419]

This fighting talk draws the lines of evolutionary argument in a very new way. For most of this century the emphasis in studies of behavior has been on establishing common ground. On demonstrating, with the help of intensive long-term field studies and sensitive analysis, that the same basic principles apply to all living things, and that humans are an integral part of this continuum. That we are naked apes, rather than fallen angels. Such work goes on, but because of the insights of a few biologists of more than usual clarity—people such as George Williams and Bill Hamilton—searching questions are now being asked about the important ways in which we do nevertheless differ, even from our nearest relatives.

The debate now concentrates on the nature of that tiny and elusive fraction of genes which we do *not* share with chimpanzees, and what having them means. It is a debate, not so much about what we are, but how we work. A gene-centered view of the world which makes it necessary for us to think about everything in terms of function and selective competition. How does this characteristic, this structure or pattern of behavior, affect our chances of survival? What does it lead us to do? And does it enhance our inclusive fitness? We may even have evolved to a point at which we can override the genetic imperatives; but if we have, then there must have been some advantage to the genes in putting us in that position. We acquired our big brains, for instance, because at the time it suited the genes that we should do so; and very little change has taken place in the structure of those brains in the 3,000 generations or so that have passed since we became indisputably sapient. There hasn't been time. But there have been changes in how we use the potential inherent in those brains, changes which depend less on biology than they do on culture. Changes which are independent of the gene in the sense that those acquired during one generation *can* now be passed on to the next—and that is a very different kind of game.

When George Bernard Shaw first encountered Darwinian thinking about evolution, he was appalled: "It seems simple, because you do not at first realize all that it involves. But when its whole significance dawns upon you, your heart sinks into a heap of sand within you. There is a hideous fatalism about it, a ghastly and damnable reduction of beauty and intelligence, of strength and purpose, of honour and aspiration."[347]

There are many who still feel that way, who accuse the new Darwinians of genetic or biological "determinism" which seems to leave no room for "free will." And they and he are right to be concerned. There is a sense in which all behavior is mediated biologically. No matter what historical influences have produced our brains, no matter what genetic instruction is encoded in them, no matter what new programs our education and experience have written there, it is still our chemical and physical organization which determines finally how we respond to any new set of stimuli. We have always been, and probably always will be, at least partly in thrall to our biology. We are only human, which is to say organisms produced by natural selection, anyway.

There is in this reality and its perception a very real danger, which reaches its limits of absurdity (one hopes) with lunacies such as the infamous "Twinkie Defense" by means of which a California lawyer convinced a jury that his client should not be convicted of murder. Junk food, he argued, had left the accused with a "diminished capacity to think clearly" and this made any premeditation of his crime impossible and meaningless. He won, and the law lost a good part of its power to establish culpability and the direct responsibility of each of us for our actions.[94]

This is no small thing. It marks a turn toward the establishment of a moral climate which attaches no blame to those who break our rules on the grounds that they are "victims of circumstance," warped by deprivation, malnutrition and social circumstance. There is good cause for considering all these influences as causal factors in rendering necessary judgment, but the risk is that each advance in biochemistry or pharmacology which links a particular chemical to a pattern of thought or behavior makes it easier to see ourselves simply as machines that react to forces over which they have no control. That may be a construct which is appealing to defense lawyers, whose sponsorship has spawned a whole new pseudo-science in which "post–traumatic stress disorder," "battered-woman syndrome," "depression-suicide compulsion" and "action-aggression addiction" play a major role. But, at heart, we know that just isn't true.

We feel this way because one of the direct results of our organic history and our chemical evolution has been a sense of justice which the genes introduced into our equation to govern reciprocal altruism. We feel that those who do good, deserve reward; and those who do bad, deserve punishment. We instinctively recognize the existence of "free will" in such things and seem to have come by this understanding, not out of conscience or compassion, but from long experience of the sort of hand-to-hand combat between individuals that is the stock in trade of natural selection.

Natural selection is simple. It makes just three assumptions. First, that individuals vary in their hereditary traits. They do. Second, that some of these traits will be better than others for ensuring reproductive success. They are. And third, that such successful traits will become more widespread in the population. They have. And that is all there is to it. Vary, multiply and let the fittest survive. "How extremely stupid not to have thought of that!" said Huxley when Darwin explained it to him.[71] But there are, of course, hidden complexities.

The result of natural selection is perhaps 30 million living species, each with its own complex gene pool. And each pool is the creation of thousands, perhaps millions, of tiny incremental advances, every one of which was an accident, something random which happened, it seems, just by trial and error. But the odds against such a long series of happy accidents are astronomical, and have spawned any number of theories about "evolution by design," which skew the odds in life's favor and push us in a particular direction. Evolutionary mathematicians point to the mass of failed experiments, of species which did not survive and have been consigned to the dustbin of genetic history, and insist that natural selection was just that, perfectly natural, and in no way acting against probability or chance. But there is also a possibility that some genetic advances could be accidents looking for a place to happen.

To the select gathering of those biologists who have challenged orthodox science in new and productive ways, I would add one last name—that of Rupert Sheldrake, a British botanist who understands natural selection as well as anyone, but who has also had the courage to suggest that the creative leaps which provide new adaptations for natural selection to work on may not be accidental at all. He maintains that patterns he calls "morphic fields" link all life, providing a pool not of genes but of successful habits that, through repetition, become more and more likely to appear by "chance." Once something has happened, it can happen again far more easily. He sees his creative fields leaping, not just from individual to individual, but from species to species across gaps of both space and time. They pass, not just from ancestors to their descendants, but also move sideways from one group of organisms to another, even if these others live on different continents. Hence the astonishing similarities between mammalian anteaters, moles, flying squirrels, cats and wolves—and the marsupial versions of each—amazing lookalikes that have evolved in isolation in Australia. And the parallel development of Old and New World porcupines, pigs and monkeys from totally different starting points.

Such "convergence" is shrugged off by most taxonomists as the result of similar environmental constraints acting on the separated species. But this is not altogether convincing and falls some way short of explaining even the more extravagant forms of mimicry and camouflage in the same habitat. I tend to agree with Sheldrake that something else is at work here in producing what he calls "evolutionary plagiarism."[348]

Sheldrake is a beguiling and sensitive man, as much poet as scientist, who shares with Darwin the capacity to outrage, while at the same time offering ideas that have a deep intuitive appeal. Darwin gave us a process of continuing organic creativity, a living and breathing idea. Sheldrake adds to that by making the idea universal. He sees his morphic fields as applying to species as well as individuals, to whole ecosystems as easily as they do to species, and perhaps even to inorganic matter in the form of entire planets and galactic systems. "It is possible," he says, "that memory is inherent in all of nature, and if we find that we are indeed living in such a world, we shall have to change our way of thinking entirely."[349]

Just so. With Sheldrake and Dawkins, out of Darwin, we keep coming back to the power of ideas as evolutionary units in their own right. No individual zebra or wildebeest inherits a route map for the awesome annual migration across the plains of Serengeti. The movement to ancestral feeding grounds along traditional paths is a pattern of group behavior, a kind of social memory that gets passed on down through all the generations. This is cultural inheritance, which has nothing to do with the genes, though it may serve their purpose by enhancing chances of survival.

Culture is, in effect, an organism taking a variety of forms, which compete against each other and change and evolve through the medium of their units of inheritance—the memes. These appear to pass from mind to mind, spreading the contagion of rumors, jokes, fads, fashions and crazes, creating short-lived mobs, longer-lasting crowds and very long-lived social or religious movements. And it may be that these, or a selection of the ones among them that have best demonstrated their inclusive fitness, are what gets embedded in us. "They are," suggested Jung, "in a sense the deposits of all our ancestral experiences."[217] Archetypes. Messages, in our case, from a million-year-old ancestor.

Sheldrake ties his ideas to Jung's by suggesting that each species stores its own collective memory in appropriate morphic fields which are drawn from the actual forms of previous similar organisms, and which help shape the nature of future members of the species. And he makes the additional suggestion that such influence takes place mostly between an organism and its *own* past states, because it is more similar to itself in the immediate past than to any other organism. In this way we keep the "idea" of ourselves going in ways that may help to explain how we heal, why we are generally symmetrical, and how we keep a recognizable form in spite of a continuous turnover of all our many chemical constituents.[350]

This sounds very much like the way in which culture is transmitted through myth and ritual in tribal society, where patterns of behavior are repeated again and again, "in the way of the ancestors," and in an unbroken line going directly back to the time such customs came into being. Ritual in all societies tends to be highly conservative, insisting that everything needs to be done in precisely the "right" way. Why? Perhaps, says Sheldrake, because such performances "really can bring the past into the present. The greater the similarity between the way the ritual is done now and the way it was done before, the stronger the resonant connection between the past and present performers of the ritual . . . The influence of past followers of such a path is explicitly recognized in many—perhaps all—religious traditions." Hence insistence on the proper use of sacred words or phrases such as prayers or mantras.[349] This is how culture is inherited or replicated, but there is nothing in such a process that allows, or can explain, the creation and origin of completely new ideas. That requires a sort of cultural mutation, with someone like Darwin or Sheldrake acting as the mutagenic agent.

Animals too get new ideas. Japanese scientists in the 1950s caught a young female macaque in the act of creating a cultural revolution which, in monkey terms, is equivalent to the invention of the wheel. She took dirty food to pools of sea water to clean and flavor it before eating, and as the human observers watched, this attractive idea took root and spread, slowly at first but with gathering momentum, through the entire island colony.[224] Something similar has been observed in Kenya where a vervet monkey, again a female, seems to have invented a new way of collecting nutritious sap by using a dry pod from an acacia tree as an absorbent tool.[179]

These primates must be given due credit for technological innovation, for setting up fully fledged cultural memes that now exist as established patterns of behavior in their communities. Less obvious, but equally striking, are social innovations which involve the intelligent and creative discovery of new uses for already established patterns. Dorothy Cheyney and Robert Seafarth in their intensive study of vervet monkeys in Kenya report at least one instance in which an adult monkey, apparently concerned about an uncontrolled fight among her relatives, succeeded in breaking it up by giving the "leopard alarm," an acoustically distinctive call which always sends all members of a troop, whatever else they may be doing on the ground, running for safety up in the trees.[68] Making such a call in the absence of a leopard is the kind of deception that will be strongly selected against. "Crying wolf" quickly devalues an alarm. But to choose to override such inhibition when that signal, and only that signal, would be effective in restoring order, is evidence of a high degree of creativity and abstraction. Something which emphasizes yet again how much of our intelligence and consciousness we owe to the stimulus provided by being part of a community of other complex creatures like ourselves.

If Sheldrake is right about morphic fields, they provide just the sort of mechanism necessary for transmission of such nongenetic information, even over large distances and across barriers of time. His notion of morphic resonance provides an explanation for uncomfortable facts such as the common readiness of some organic fluids to crystallize readily anywhere, once such behavior has been coaxed somewhere out of one sample of a substance that has stubbornly refused to take its solid form despite years of strenuous effort. But even if his hypothesis can be shown to be faulty, it is clear that in our species at least we can transmit such information by direct cultural contact alone. We have, to a very large extent, liberated ourselves from strict genetic control and put a rival pattern of inheritance in play. We have completed the first step in the campaign incited by Huxley and Williams, and taken the battle directly to the enemy.

This "enemy," remember, is the same process of natural selection that has resulted in the creation of birds of paradise and the bottle-nosed dolphin. It is the very pattern which set in train the sequence of events that led to our ancestors becoming bipedal and pushy. It is a power which continues to shape the living world around us in ways that are elegant, appropriate and ecologically sound. It is the fount of all that is "just right" and awesomely beautiful in nature. It is all these things, but it is also morally and ethically unsound.

Natural selection is extraordinarily good at maximizing immediate genetic interests, but it is uncommonly bad at long-term planning. Our versatile grasping hand with its opposable thumb and fluid rotation is a subtle piece of bioengineering, apparently perfected by natural selection for the sort of precise manipulation that has made us human. But, as George Williams points out, such selection could only act on an organ already adapted to hold an overhead tree branch while turning the body around that axis. "The production of an organ capable of threading needles and turning doorknobs was made possible as an incidental consequence of selection to be better than one's neighbor at swinging through the trees. The helping hand of the Good Samaritan and the motivation for its use, raise no question on the malice or power of natural selection. They merely show that this persistent and powerful enemy is a mindless fool."[420]

The same unthinking process has also succeeded in producing apparently unselfish altruistic acts. We can't help but be impressed when several childless female langurs come to the aid of a female whose infant a new group leader is trying to kill, but it is as well to look at the behavior in sociobiological terms before leaping to conclusions about its apparent ethical content. Generally speaking, if one organism ever does anything to help another it is bound to be either an accident; the result of manipulative deception; a nepotistic act in favor of a close relative; or one of reciprocity that sooner or later will prove equally profitable to the giver.

Charity doesn't come naturally to us. It is the result of deliberate choice, of a conscious decision to revolt against the tyranny of the genes. It comes from an awareness of difference that is comparatively recent and has made it necessary for us to admit, along with philosopher Peter Singer, that: "We do not find our ethical premises in our biological nature, or under cabbages either. We choose them." And it is that act of choice, a decision against nature, that has set a brand-new and feisty cat in amongst our moral pigeons.

We were social long before we became human. And in that long social experience lie the biological origins of virtues such as compassion, empathy, love, conscience and a powerful sense of justice. All these things now have a firm genetic base and could be seen as natural moral values. "That's the good news," says Robert Wright in an impressive new look at the science of evolutionary psychology. "The bad news is that, although these things are in some ways blessings for humanity as a whole, they didn't evolve for the 'good of the species' and aren't reliably employed to that end." We switch them on and off as it suits us and, thanks to genetic pressures, do so even without thinking. "Human beings are a species splendid in their array of moral equipment, tragic in their propensity to misuse it, and pathetic in their constitutional ignorance of the misuse."[433]

The net result of this confusion is conflict between an old set of impulses which are, by design, very strong; and a new set of values which are, inevitably, unnatural. If we choose to be on the side of the angels, then the enemy is on the side of the genes, but the lines are far from clearly drawn. Those same genes and their selfish interests also succeed in creating everything in our nature that appears to be benign. But it is beginning to seem likely that the roots of all we now regard as evil, weak or strong, lie very firmly in the camp, and in the action, of natural selection. And we must expect to find that attitudes which arise directly from biology will be working in the enemy's service, not in ours.

The one weapon that is unquestionably ours is human speech, and the fossils it leaves in the form of the written or printed word. With these, we have a system of information transmission which is a real rival to genetic reproduction. Cultural evolution is very much like genetic evolution in that progress can be tracked by the effects each has on individuals who receive them. Memes rise and fall in a population exactly as genes do, spreading in an environment that is favorable to them, slipping into extinction where they fail to find such favor. New ideas, new music, new foods, new fashions and new faiths all succeed or fail on their merits. The difference is that, unlike the genes, they do not necessarily have to contribute to fitness in any way. Celibacy, hard rock, sushi, nipple rings and walking over hot coals have no obvious survival value, except as stimuli that can keep big brains busy and amused. At least for a while. They are not subject to natural selection, but are liable to be lost as soon as they fail to satisfy an equally natural human curiosity. They are also likely to succeed beyond all dreams of avarice if they have been canny enough to have a foot in both camps, trading on novelty value as well as a biological compulsion. The meme for inhaling nicotine, for instance, was always going to be a winner, despite its lethal consequences.

Happily, however, there are also fugitives from the genetic camp that aid the cultural cause. Skill in deception is a natural talent, the logical outcome of a tendency to conflate and confuse as a way of gaining genetic advantage. But the equally logical counter to this tissue of lies is an extremely fine sense for the detection of such fraud in others. We never stop doing it. Much of our time is spent in what Wright calls "the unending, unconscious, all-embracing search for signs of unworthiness," that has now become so deeply entrenched that we have to make a conscious effort to control it. It has become all too easy to point out the faults in others. But the fact that we see them so well, and never stop looking, does have considerable survival value. It provides a useful antidote to the blotting-paper stage of enculturation, in which we absorb huge and uncritical masses of raw information. The tendency of the curious child to believe everything it is told is followed directly and properly by the tendency of the rebellious teenager to believe nothing at all. Later, if we are lucky, we settle into a sensible adult compromise between the two extremes.

If there is one part of the child's wide-eyed wonder at everything that does survive, it is that which makes it possible for us to take some matters—cultures tend to differ about which ones—on faith. Every group of people stubbornly holds on to at least one area of life in which we are prepared to suspend disbelief and not ask for objective assessment or proof. Almost everyone has some kind of religion.

Jane Goodall describes a heavy thunderstorm at Gombe which provoked four male chimps to charge up and down a hillside, tearing off branches, hooting wildly and drumming against tree trunks, thrashing and dancing in the rain as the storm raged around them.[151] There is in this defiance, this shaking of fists at the heavens, a hint of the origin of religion. A sort of drumming in the dark that so excited and simultaneously soothed me as a child in

Africa, when the last oil lamp in our home had been extinguished and we were embraced by the kind of darkness one rarely sees any more. The options then didn't close down as the senses were eliminated one by one; they actually increased with each subtraction, until the drumming in a distant kraal was as real and as close to me as it was for the drummer himself. We drummed and I felt dilated, part of something very old and very large, something total that needed no measurement to verify its reality.

We are born animists, happy to believe that everything we encounter is alive, just as *we* are, and that all objects are equally able to encounter us, for good or evil. It is a fortunate disposition, a time in both personal and human history when we are closest to nature, most easily touched by mystical experience, most accessible to a connection that provides a real sense of the presence of power around us. It lasts only as long as we let it, beginning to slip away as soon as we become more demanding, more distant, more inclined to ask inappropriate questions that push the presence away, like a dream that refuses to be brought to conscious attention. And that is when we become more involved with organized religion, more inclined to fall into history and to look for heroes and redeemers who can meet the numinous, walking and talking with gods on our behalf. From then on, "The world is too much with us," and we are, in all the ways that matter, out of tune.[436]

In science, our heroes are Galileo, Newton and Darwin, and the battles they have fought and won have been epic, both in scale and in consequence. They leave us, however, in something of a bind. We have a far better idea of how the machinery of the world works, but we still don't know what it is expected to do, or if there is any grand plan at all. Religions have a vested interest in an afterlife as part of the big picture. How else do you sell a meme that demands more self-restraint in this one? But their prescriptions for the good life here vary enough to produce both confusion and dissension. The one thing, perhaps the only thing, on which all religions seem to agree is demonic temptation.

There always seems to be an evil being who tries to entice us into doing wrong. In the Bible and Koran, it is Satan; while in Buddhist scripture he appears as the arch-tempter Mara. Their job description is the same: "Look innocent and lure humans into minor transgressions that can grow into major problems." Start with a little gentle flirting and let things escalate from there into full-scale orgies. Robert Wright suggests that: "It is no coincidence that demons and drug dealers often use the same opening line (Just try a little, it will feel good), or that religious people often see demons in drugs." Though he adds, parenthetically, that: "One habit that natural selection never 'meant' to encourage was drug-addiction itself. This miracle of technology is an unanticipated biochemical intervention, a subversion of the reward system. We were meant to get our thrills the old-fashioned way."[433] And we can perhaps blame some kinds of addiction, not just on the drugs, but on infection by a meme that insists that injecting heroin is fun. In time that meme at least seems likely to be selected against, if not by common sense, then at least by its own lethal consequences.

In all religions, the tempter is malign, a dark and belligerent force that has to be combated. The battle itself is always inevitable, a given; and training for it invariably takes the form of abstinence of one sort or another from earthly pleasures such as eating, drinking, making love or undermining rivals. What is expected of converts is nothing more nor less than full defiance of human nature, which is usually described as evil—evidence of original sin. The advice of all the sages is to rebel against the forces that seek to control us by their unholy system of values; to refuse to play their game. And stripped of all the mythic trim, the regional color and demonic personification, what this clarion call amounts to is a struggle against the selfish interests of the gene.

It is the genes that "want" men to have sex with as many women as possible. Sex and power are the nuts and bolts of natural selection. We are designed to seek ephemeral pleasures, to believe that just one more million, the next lover, one last drink for the road, will bring everlasting bliss. "Natural selection," says Wright, "has a malicious sense of humor; it leads us along with a series of promises and then keeps saying 'Just kidding'."[433] And the force which drives us into illicit pleasures and hides behind the guises of temptation is clearly identifiable now as one whose interests are served by practices which may no longer be appropriate for us. It is the devil we know. Our old ally, the selfish gene.

The ongoing argument about nature and nurture, about inheritance and learning, genes and culture, is no longer one just of academic interest. It has become personal, a divide with which every one of us has to deal in our own lives. And for reasons not unconnected with the hellfire doctrines of old-fashioned Presbyterianism, many of those who seem to have come closest to understanding the dilemma and putting it directly into starkly simple black and white terms have been Scots.

Robert Louis Stevenson was also a romantic. A frail tubercular body with a swashbuckling soul, who saw life clearly enough, but was determined to make an adventure of it. He did and died, as he would have liked, as a patriarch in the South Seas, where he was known as Tusitala—"the teller of tales." As a young man, Stevenson admired, and wrote his first book about, the Covenanters, a group of strict Presbyterians whose struggles for religious liberty led to the English Civil War, several rebellions and their own persecution in the seventeenth century. Stevenson himself, however, soon rebelled against the constraints of his parents' religion and set himself up as a liberal Bohemian, but the memory lingered because in 1886 he set down a simple moral tale which stands still as the most vivid and beguiling portrait of our divided selves.

"I had long been trying to write a story on the subject," Stevenson said later, "to find a body, a vehicle, for that strong sense of man's double being which must at times come in upon and overwhelm the mind of every thinking creature."[366] And the inspiration finally came to him in a dream that produced all the basic plot elements for *The Strange Case of Dr. Jekyll and Mr. Hyde.*[365]

Henry Jekyll is torn, as we all are, between his "good" self and his "bad" desires. He calls it his "dual nature" and describes how "both sides of me were in dead earnest; I was no more myself when I laid aside restraint and plunged in shame, than when I labored, in the eye of day, at the furtherance of knowledge or the relief of sorrow and suffering." Man, he decides, "is not truly one, but truly two," the product of a "thorough and primitive duality."

So far so good. That is the human condition. But then the good

Dr. Jekyll makes his first mistake. He plans to separate the two identities, in the hope that "life would be relieved of all that was unbearable," and succeeds, with the help of a drug, in doing just that. A little white powder, and Edward Hyde appears—smaller, younger and darker, with "a haunting sense of unexpressed deformity"—free to indulge in all those pleasures which Jekyll had denied himself. Hyde moves into a separate apartment and pursues interests which at first are just "undignified," but which soon "turn towards the monstrous." But Jekyll feels no guilt because: "It was Hyde, after all, and Hyde alone that was guilty. Jekyll was no worse; he woke again to his good qualities seemingly unimpaired; he would even make haste, where it was possible, to undo the evil done by Hyde. And thus conscience slumbered."

But the plot thickened. Hyde started killing and Jekyll needed more and more of the drug to regain himself and his composure. Yet he continued with the experiment, certain that: "The moment I choose, I can be rid of Mr. Hyde. I give you my hand upon that." And that was Dr. Jekyll's second and fatal mistake, the hubris of assuming that it was possible for one part of his nature to do without the other. Before long, he starts turning into Hyde without the drug and ends up confined to his laboratory, where eventually he takes poison and ends both of their lives.

Stevenson weaves a spellbinding tale which on the surface is little more than another Gothic horror story in which a prim person tries to have fun in disguise. But his little moral allegory soon turns into a monumental study of the origins and nature of human evil. Stevenson draws on notions of duality already explored by Saint Augustine and Thomas Aquinas, but brings the concept so close to home that even those who have never read the book know and accept its premise. It is a story which grips imagination in the way only archetypes, which have a universal significance, can. And it took Stevenson just six days to put together something so insightful and perceptive that it anticipates Freud and Jung's writings on the subject by decades.[333]

Stevenson seems also to have understood, at least intuitively, the biological nature of the dilemma. He makes Hyde not so much cruel, but callous, acting without any hint of normal human sympathy. Our first glimpse of the alter ego is one of chilling inhumanity. Hyde knocked down a young girl by accident at a street corner and then "trampled calmly over the child's body and left her screaming on the ground. It sounds nothing to hear, but it was hellish to see." Where there ought to be empathy, there is an awful vacuum. Jekyll worries about Hyde: "I do sincerely take a very great interest in that young man." But Hyde is indifferent to Jekyll, "or but remembered him as the mountain bandit remembers the cavern in which he conceals himself from pursuit. Jekyll had more than a father's interest (because he shared Hyde's pleasures): Hyde had more than a son's indifference."

It is hard to imagine a better description of the gene's relationship with the body it occupies, and even more difficult to conjure up an image that suggests both the power and the callousness of the Other with such clarity. In his final letter, which tells the whole story from his point of view, Henry Jekyll describes Edward Hyde's: "wonderful selfishness and circumscription to the moment" in a way that perfectly catches the feeling of genetic self-absorption and short-sighted concern. And in Stevenson's description of Hyde killing: "With a transport of glee . . . tasting delight with every blow," there is a terrible prescience of the sort of behavior shown in the same circumstances, possibly under the same emotionally crippled control, by children such as Mary Bell, Robert Thompson and Eric Smith, who laugh even as they kill.

Biologist Robert Trivers is probably right in suggesting that friendship, gratitude, sympathy, guilt, a sense of fairness or justice, and even the rudiments of morality, are apparent in our experience of reciprocal altruism.[388] They have their roots in genetic selfishness, but we should not make the Jekyllian mistake of assuming that now we have them, we can dispense with genetic participation altogether. Like it or not, we are locked into a relationship that stands or falls as such. As a double. We and our indispensable shadow. Literature is full of warnings about those who have tried to shake it. Peter Schlemihl sold his to the devil and lived to regret it. Dorian Gray let his portrait take all the blame, but it got him in the end. Hans Christian Andersen tells a story about a dismissed shadow that comes back with a new body that reduces the former owner to the status of its shadow—and then kills him. It seems that you can't be too careful about such things.[381]

It was Jung who eventually gave the shadow proper substance. "How can I be substantial," he asked, "if I fail to cast a shadow? I must have a dark side also if I am to be whole; and inasmuch as I become conscious of my shadow I also remember that I am a human being like any other."[215] At first he described the shadow in Freudian terms, simply as a repressed part of personality, but by 1917 he had upgraded it to the status of a separate unconscious personality in its own right, something that embarrasses us and holds all the unpleasant qualities we like to hide. The "Other" in us that needs to be acknowledged, and is even more terrible if unacknowledged or ignored.[214]

Dr. Frankenstein, in Mary Shelley's astonishing allegory, succeeds in creating new life. But when he finds that it is not the perfect creature he had in mind, rejects it and disclaims all responsibility for any of its defects. He repudiates his own shadow, and from that moment everything goes wrong. The monster becomes resentful and destructive and sets out on a trail of havoc that ends in the inevitable destruction of them both. Like Jekyll and Hyde, the duo of Frankenstein and his monster—who have become so entangled in our minds that one is often confused with the other—are now part of our mythology. Their story is more than just a cautionary tale about scientific hubris. It speaks as directly to us now, about good and evil and the integration of personality, as it did in the summer of 1816 when Mary scared herself with her dream of "the pale student of unhallowed arts kneeling beside the thing he had put together."[352]

Her story works so well because it fits in with what we feel about ourselves and about our desires, providing confirmation of an internal split that amounts to moral meiosis. We suspect that we all have the potential to be destructive and hope that, when the time comes, we can keep such tendencies under control. There is solace to be found in laying the blame elsewhere, on a Jekyll or a Frankenstein or even God, but if there really is "something wicked" in us, it makes more sense to see it as our problem, not theirs. Job learned that. He did not deserve the miseries to which he was subjected—there seemed no point to them and he said so—but he was denied even death as a solution. There are no easy answers. The stark reality of that Hebraic tradition was softened by Christian theologians, who made much of Job's role as a foot-soldier of God in his ongoing battle against the devil. But the harder truth is a lot closer to home. It's not the devil who is to blame, but that part of us that operates its own agenda, ruthlessly ignoring the possibility of suffering and very real pain for the rest.

The moral contained in the Book of Job is that natural authority (in this case an unusually vengeful God) can be tyrannical and overpowering. It is an overwhelming force, but also a blind one that is too unconscious to be moral. Sound familiar? It should, because that is a good description of how the genes exercise their authority. In his *Answer to Job*, Jung says that the prophet is in this case morally superior to God, who acts as though "might is right." The morality has to be supplied instead by Job, as it must be supplied by us in all our dealings with the genetic government.[216]

I am conscious at this delicate point of circumventing all the long arguments of theodicy in a somewhat cavalier way. Saint Augustine, Irenaeus the Bishop of Lyons, and more recently the theologian John Hick, have all labored to produce a standard Christian answer to the existence of evil in a universe designed and presided over by a good God.[183] Their musings have, to my mind, been superseded even in my lifetime by events in Nazi Germany, Chile, Cambodia, Uganda, Vietnam, Serbia and Rwanda, to mention but a few: which make it clear that we, and we alone, bear the blame. History, even in this century, has confronted us squarely with our own demonic capacities. We have, like Faust finally coming face to face with Mephistopheles, been forced to concede that the mask he wears bears features very much like our own. He is us, and neither evil nor we can be redeemed. We just are; and we very clearly are the products, for good or evil, of our biological evolution.

A Polish guard called as a witness before the Tribunal at the Nuremberg Trials gave this testimony:

> Women carrying children were always sent with them to the crematorium. Children were of no labor value, so they were

> killed. The mothers were sent along too, because separation might lead to panic and hysteria, which might slow up the destruction process, and this could not be afforded. It was simpler to kill the mothers too, and keep things quiet and smooth. The children were torn from their parents outside the crematorium and sent to the gas chambers separately. At that point, crowding more people into the chambers became the most urgent consideration. Separating meant that more children could be packed in separately, or they could be thrown in over the heads of the adults once the chambers were packed. When the extermination of the Jews in the gas chambers was at its height, orders were issued that children were to be thrown straight into the crematorium furnaces, or into a pit near the crematorium, without being gassed first.[370]

There are few bleaker statements to be found anywhere in human history, and no possibility of justifying such clinical brutality as part of any divine plan—even one stretched beyond imagining to some distant future when all this can be seen to have been necessary. But there are ready, if scarcely less sanguine explanations, available in biology. The first genetic rule, remember, is: "Be nasty to outsiders." And everything had been done in Germany during the 1930s to separate and identify and classify Jews not just as outsiders, but as outsiders with evil intent. Whatever happened to them was not only sanctioned by the state, but was beyond genetic as well as legal reproach on the grounds that Jews were less than human. So German commanders and Polish guards in the death camps, nice men who loved their children and were kind to their dogs, people who were probably regular churchgoers and believed in God, became involved in infamy.

"We were just following genetic orders," was never offered up in defense at Nuremberg. It does have some merit as a mitigating factor, but was never even considered for the very good reason that international law and our own hearts tell us that it is no longer a valid excuse. As civilized human beings, we are expected to know better, not out of fear of subsequent punishment, but simply because an informal moral code now exists in which there is an implicit understanding that we ought to be able by now to do better than the genetic minimum. Human nature is no longer inevitable. As Katharine Hepburn says to Humphrey Bogart in *The African Queen:* "Nature, Mister Allnutt, is what we are put in this world to rise above."

Quite so, though for the sake of dialectic simplicity, I would prefer to drop the "put" from her wonderful put-down. That is part of another argument. The genes have only themselves to blame for our being here. What matters now is that the best we can do with our genetic programming, once it has done its job of getting us to the starting line, is to use it as a model of how not to run the race, or the species. "A good starting point," suggests Robert Wright, "would be to generally discount moral indignation by 50 percent or so, mindful of its built-in bias, and to be similarly suspicious of moral indifference to suffering."[433] We need to be especially vigilant in all situations where the interests of our group—our family, clan, tribe or nation—are at odds with those of another group of ready-made outsiders. And we have to be constantly aware of genetic tendencies to be too hard on lower-status individuals; and far too tolerant of those whose position, wealth or beauty lead us to grant them special privileges. That isn't necessary now. We are not just monkeys any more. But we are still primates, with all the baggage that implies.

Status continues to be important to us. It matters how others see us. High public esteem can provide enormous genetic rewards, and low esteem banish us to the reproductive equivalent of Siberia. Status symbols are vital, and a certain amount of conspicuous consumption and verbal self-inflation are to be expected. But not too much. Those who manage to get the balance just right walk the same fine line as up-and-coming young males in a baboon colony have to tread. Every increase in your own status is one gained at the expense of someone else. And every step in such a climb has to be negotiated with the kind of political flair that both troop leaders and company presidents understand very well. This sort of social sensitivity is a vital part of our genetic inheritance, the part we are going to have to cultivate to find a sensible balance between the rival forces of genetic and cultural evolution. But, at the moment, we are not doing very well.

Here, on the brink of a new millennium, we find ourselves in dire straits. At the mercy of genetic forces which evolved to meet situations and to fit social patterns that no longer exist. We are moving rapidly to meet what writer Robert Kaplan describes vividly as "The Coming Anarchy." Our world, he points out, is changing in ways that make nonsense of maps and nation-states. The true divisions lie now not between clans or tribes or countries, but between a small, healthy, well-fed minority who see population growth as a threat to their quality of life; and the rest of our species for whom large families present the only hope of escape from a way of life that is truly Hobbesian.

We have moved from nation-state conflict, through ideological conflict into a post-Cold War era of cultural conflict. The cities keep growing, coalescing with one another into megalopolitan sprawls and endless shanty towns, where sixty percent of all the world's people will soon struggle through streets paved with garbage and irrigated by open sewers. While in the country, deforestation, soil erosion, water depletion and chemical pollution produce shortages that are already leading to something very like anarchy.[221]

In this environment, existing hatreds become inflamed; social, economic and religious motives become hopelessly entangled; borders cease to have any meaning; new epidemics break out and old diseases turn once again into plagues. One day, if we survive, there may be a new, racially hybridized, culturally homogenized, global society. That could be our only hope—to become genetically similar enough to care a little more for one another—though I view such a prospect with dismay, because it falls under the pathic principle that real order is destroyed, rather than created, by a loss of diversity.[136]

We need to learn to love our differences. But in the meantime, and for a long time, we are going to be faced with chaos and political disintegration—both at levels which will make the ideal of democratic capitalism seem laughable. The process has already begun with ethnic violence, unprovoked crime, ruthless regimes and the migration of millions of permanent refugees. Central governments have begun to wither, regional and tribal identities are being revived, and fundamentalism of every stripe is becoming ever more obvious and more strident. Under the pressure of environmental and demographic stress, we are watching the old order crack and fragment into city-states, shanty-states and ragtag private armies belonging to a whole new generation of local warlords. Authority in many areas no longer rests in government or military hands, but has devolved upon Serbian "chetniks" and Somalian "technicals" who are, understandably, impervious to the appeals of international or intergovernmental organizations.

In evolutionary terms, what we are seeing is the rebirth of warrior societies, the reprivatizing of mankind into smaller, more ruthless communities in which the radius of trust is being limited once again to immediate kin and close comrades. We are being retribalized, closing the genetic gates; but with nature breaking down and clean water in dangerously short supply, our biological roots can no longer provide us with the moral nourishment we need.

That is what makes diversity so valuable and so resilient. Without it, we become involved in a brutal new process of natural selection by the most unnatural of means. Virtue in this kind of endeavor will come to lie only in that which provides personal security or tenuous short-term communal advantage. There will no longer be any way of separating war and crime. The distribution has already become terminally blurred by wars fought for and by criminal interests such as the drug cartels. There will be no possibility any more of honoring the Augustinian conditions of just cause, proper authority and right intention. There will, if this dreadful scenario is allowed to play itself out, be no health in us. Just undiluted, all-encompassing evil.

As I write this, Rwanda moves relentlessly from holocaust to hunger, its images flickering through my mind like scenes from an inferno. A woman dances, naked, down a dusty highway, cursing at the listless crowds and at corpses lying by the roadside ... A man at the edge of a mass grave laughs in delight when he manages to toss the lifeless body of a child squarely into the middle of the pit ... Another child lying among the corpses rolls over, still breathing, but nobody goes to its aid . . . A group of men fall upon a passerby, beat him senseless, then lay him on his stomach and stomp on his spine until it breaks with an audible snap. These are the stuff of nightmares and nothing I can do or say will make them go away. This is Armageddon with commercial breaks, evidence that the demons are not just among us—they are us. Or at least that part of us that lashes out from limbo, festering in the place between what was, and that which may yet come to be.

Statisticians tell us that the figures for violent crime have remained flat over the past twenty years, and that it is only our perception of crime which has changed. Perhaps, but this sort of reassurance feels as phony as meteorological insistence that the climate isn't really changing either. Those out there on the front—farmers, fishermen and policemen on foot patrol—know that it all feels very different. You used to be able to leave your home unlocked or your horse unguarded in its field. Now they come in and defecate on your carpet, and sexually mutilate the animal. A fine Welsh cob mare, four weeks off foaling, had her womb slashed open from the inside by someone with a blade strapped to a pole, in just one of a frenzy of horse-ripping incidents that outraged Hampshire in England in 1993.[144] The product of a meme sown by Peter Shaffer twenty years earlier in his play *Equus,* in which a deranged stable boy blinds a horse after it has witnessed his first unsuccessful sexual encounter? Who knows, but there is little doubt that the moral climate is changing.

Our current fear of violence is so great that it threatens to skew our judgment altogether. Recent news from the United States suggests that even boyhood is coming to be seen as a pathological condition. "Does the boy disregard authority, make snide remarks in class, push other kids around and play hooky? Maybe he has a conduct disorder. Is he fidgety, impulsive, disruptive, easily bored? Perhaps he is suffering from attention-deficit hyperactivity disorder, the disease of the hour. Does he prefer computer games and goofing off to homework? He might have dyslexia."[10] The drug of choice for curbing such waywardness is the depressant Ritalin, which seems likely now to be prescribed, retrospectively, for even Tom Sawyer and Huckleberry Finn.

Which is all ludicrous, of course. But we do seem to favor extreme reactions, particularly toward any individuals who can be lumped together in our minds as a convenient "group" to be dehumanized and identified with characteristics of which we disapprove. This seems to be one of the reasons why violence has exploded to the point where happiness has come to mean a warm gun, preferably fresh from use against one of "them"—the gays, the blacks, the liberals, the abortionists, the vegetarians—you know, the ones who hate "us." It is high time we recognized such genetic conceits for exactly what they are: self-serving, once useful, now nonsensical. Groups are not moral or immoral. Morality is a personal matter, something that can be acquired by, and attributed only to, individuals. Evolution and natural selection happen to us all, but just one at a time.

And yet, as we have seen, reciprocal altruism is something that natural selection favors. In some respects, evolution has been bringing us closer together, quite automatically and unconsciously, encouraging giving and receiving. The result, as George Williams points out, is that: "An individual who maximises his friendships and minimises his antagonisms will have an evolutionary advantage, and selection should favour those characters that promote the optimisation of personal relationships."[418] This is one of the happier, quite unintended, results of genetic selfishness and Tit For Tat. One of those wonderful ways in which things slide into beauty and balance without even trying.

In human terms, Tit For Tat is seldom as simple and binary as game theorists would like. At the start of a relationship, we might trade one-for-one, giving and receiving alternate invitations to dinner; but friends soon let down their guard with one another, and only once in a while review their joint account, monitoring perhaps just for signs of incipient meanness. This is the sort of useful social cement we have inherited from a genetic process that begins with random mutations, with accidents made meaningful by natural selection. Simple day-to-day competition results in the accumulation, not just of genetic benefits, but rewards which touch our lives by enhancing social harmony—even among individuals who have no genetic relationship at all.

And that is the key. The magic factor which makes it possible for us all to rise above nature in ways that would have delighted Rosie Sayer on the *African Queen.* We are not naturally moral, but just recognizing that fact is the essential first step in learning how to become so. The good news is that kin selection—being nice to insiders; and guilt—a reminder not to be nasty to insiders—coupled with the power of reciprocal altruism, are creating an expanding web of trust, obligation and affection. And will continue to do so, given half a chance.

The problem is that love is *not* all you need. Unconditional love is a recipe for disaster. It doesn't work unless everyone is doing it. And in a society where no one gets punished for anything, immoral behavior blossoms and grows. Putting aside, for the moment, the whole question of the morality of punishment and the power of retribution, the only pragmatic solution to selfishly competitive behavior is one the genes will easily understand. Take the profit out of it. The genes may be mindless, but they are not stupid. If a behavior offers no benefits, the genes will very soon drop it and push evolution in some other direction, one that *does* offer a selective advantage. That is how and why the Tit For Tat strategy works. It holds individuals responsible for their actions, but only as long as such accountability actually does the system good.

We come back, with this notion, to where my whole quest began—to the Aristotelean idea of "just enough", to the Goldilocks solution and the notion that evil can be defined as anything that upsets the ecology. Anything that works against, rather than for, equilibrium.

The nineteenth-century philosopher John Stuart Mill, in his usual lucid way, said that morality was simple. Something is "good" if it raises the amount of happiness in the world, "bad" if it increases suffering. His doctrine of *utilitarianism* boils down to the minimal assertion that, all other things being equal, happiness is more desirable than unhappiness; and that morality is the code that leads most directly to the greater happiness. Given the wide cultural variation in what exactly seems to make people happy, this general belief in the "goodness" of happiness and the "badness" of suffering may be the only common moral ground on which all humans can meet. But, happily, it is one which translates easily and usefully into biological terms.

If I pass up the chance to cheat you, and you respond by being equally nice to me, as Tit For Tat requires, then we are both better off. The alternative of constantly cheating on one another is a game both lose in the end, because it becomes very costly in terms of increased fear and vigilance. But if we can, by our example, encourage others to extend similar courtesies to those around them—which is all any moral code seeks to do—we contribute directly to the greater happiness of the whole system, and that has to be good for the ecology. But there is a problem.

Contrary to Mill's liberal assertion that everyone's happiness counts equally, there stands a very basic genetic assumption that some individuals *are* more equal than others. *My* happiness is more important than yours. "Leave aside for the moment that pursuing goals which promise to make us happy, in the long run, often doesn't," suggests Robert Wright. "Leave aside that natural selection doesn't really 'care' about our happiness in the end and will readily countenance our suffering if that will get our genes into the next generation. For now, the point is that the basic mechanism by which our genes control us is the deep, often unspoken (even unthought), conviction that our happiness is special. We are designed not to worry about anyone else's happiness, except in the sort of cases where such worrying has, during evolution, benefited our genes."[433]

Everything, everywhere, works like that. Genetic evolution is the ultimate in selfishness. All three genetic laws—be nice to insiders, nasty to outsiders, and cheat a lot—require it. The behavior they encourage is *designed* to interfere with the happiness of others. If genes were capable of having emotions, chromosomal samplers would say things like: "Happiness is making others unhappy." Nature may succeed in being astonishingly beautiful, but we should not expect it to have any moral authority. On the contrary, we need to keep reminding ourselves that Huxley's "enemy" and ours is one whose strengths depend upon billions of years of selection *for* selfishness.

Knowing this should make us suspicious, for a start, of any attitudes or opinions which are based on family traditions or group loyalties. These are honor-bound to be based on narrow genetic self-interest. Genetic morality is the biological equivalent of original sin—something we need to overcome. And our best weapon in this battle is the very one the genes themselves forced into existence by their tendency to cheat at every possible opportunity. Suspicion of the motives of others is a vital counterploy; and when it is combined with reason, when we can think through the consequences to us of complying with or defying genetic propaganda, then we get a taste of real free will.

It is all too easy, however, to become confused about which instructions are genetic and which arise anew from human reason. And such confusion can be tragic. In two recent instances, it has even led to mass murder.

On March 20th, 1995, twelve people died and over five thousand others were sickened or hospitalized following a multiple gas attack on Tokyo's subway system. The goal, apparently, was to destroy the Kasumagaseki section of the city, in which most government offices are housed. And the weapon was ten liters of a nerve gas called sarin, manufactured for the purpose by a mysterious organization based at the foot of Mount Fuji.

Aum Shinrikyo, which means "Supreme Truth," is a doomsday cult led by a blind guru called Shoko Asahara who claims to be a reincarnation of the Buddha. His millennial prophecies have attracted at least thirty thousand followers in Japan and given the group assets of over a billion U.S. dollars, more than enough money to fund a series of munitions factories and chemical laboratories. Scientists and technicians in these facilities at Aum produced enough sarin to kill 8 million people and were in the process of developing small remote-controlled helicopters to deliver the gas in devastating mass attacks on major Japanese cities in November of 1995.

This assault was to be augmented by germ warfare, including the release of the Ebola virus, and blamed on the United States of America, which Aum believed was plotting anyway to take over Japan in 1997. The cultists were determined to survive such an invasion and had set up assembly lines capable of producing their own automatic weapons and even tanks, when their "Supreme Master" changed his plans and decided that it was necessary not just to defend themselves, but to carry out a preemptive strike. One shocking enough perhaps even to start a world war.

In April of 1995, an attempt was made to assassinate the Tokyo police chief, and on May 5th a cyanide attack was mounted on Shinjuku Station. But by May 16th, the cult leader was in custody along with 34 of his followers, all charged with murder or the conspiracy to commit murder—not just in the Tokyo subway, but in a gas attack in Matsumoto the previous year, which killed seven people; in the disappearance in 1989 of a lawyer who was investigating Aum; and in the ritual deaths of scores of cult members whose remains appear to have been disposed of in a cement-grinding machine.

And these horrors have been mirrored halfway round the world by other groups of well-armed outcasts digging in for their own grisly Apocalypse.

On April 19th, 1995, a rented truck carrying 4,000 pounds of explosives was detonated in front of a federal building in the center of Oklahoma City. The blast, which was felt thirty miles away, all but demolished the nine-story structure, killing 167 people and injuring hundreds of others.

This attack was mounted, it seems, by a small group of disaffected right-wingers led by Timothy McVeigh, a 27-year-old ex-soldier obsessed by fear and hatred of his own government. He, and many others like him, feed on conspiracy theories which see the Attorney General of the United States as a paid agent of Jewish Colombian drug lords, and AIDS as the product of secret laboratories involved in the organized destruction of political dissidents. They believe that Russian troops remain hidden in salt mines beneath Detroit, and that the United Nations, in collusion with Los Angeles street gangs, has a sinister plan to disarm and subjugate the American people.

Some of these zealots are organized into well-funded militias with names that, like the Supreme Truth, appeal to patriotism and populist dissatisfaction. They claim to be defenders of liberty, justice, Christianity, the Constitution and white supremacy. And, almost without exception, they favor a heavily armed response to what they perceive as "The Enemy Within."

Not all these "true believers" run around in camouflage or pink pyjama suits, and they may not worry about the same things. The Michigan Militia Corps sent inflammatory fax messages to the news media, blaming the Oklahoma explosion on the Japanese government. But there is a common denominator which links the Guardians of American Liberty, the Aryan Nations and Aum Shinrikyo. They are all paranoid.

It makes sense to be sceptical. We need to learn how to think for ourselves. How to question some of the "truths" we pick up in the blotting-paper phase of our early lives. But it is alarmingly easy to let disillusion set in and to allow disappointments, slights, misunderstandings and dozens of little humiliations to accumulate. These are part of the baggage involved in becoming adult and need to be shed, along with ideas that don't work or cannot be squared with an emerging sense of independence and fair play. We have to reconcile inherent selfishness and required generosity. And that isn't easy.

Psychoanalyst Melanie Klein showed how vivid children's fantasy lives can be, and how often innate aggressiveness can be projected onto a mother whose difficult task it is to both provide and deny. The result, more often than not, is ambivalence—a condition in which love and hate, patriotism and treason, coexist and can lead to both denial and delusion. And it is a small step from necessary and inevitable neurosis, which we all know well and in which our sanity is never in doubt; to fully-fledged functional psychosis, in which we are quite clearly no longer in our right minds.

Paranoia is characterized by touchy and suspicious behavior that can grow into delusions of persecution. Typically, these do not involve intellectual deterioration. Paranoids organize their delusions into coherent, internally consistent systems on which they are prepared to act, even to kill. Such delusions can be as farcical, and truly deluded, as the idea that the purchase of Burger King by a British company proves that Queen Elizabeth is behind a plot to regain control of the American colonies. But rehearsal of such wild ideas among others with similar delusions can keep old wounds open long enough to feed incipient paranoia and turn it into a lethal grievance, not just against a mother or a mother-country, but against all forms of authority. And it is clear that paranoia induces further paranoia.

Just five days after the Oklahoma outrage, the president of the California Forestry Association in Sacramento received the last of a series of carefully crafted and deadly packages that ended up killing three people and injuring another twenty-two. All the victims were strangers to the killer, an intellectual hermit who nursed an imaginary grievance against industrial society and modern civilization.

It took eighteen years to track down Theodore John Kaczynski. But what succeeded in the end was not so much an obsessive manhunt, one of the longest and largest in FBI history, but the Unabomber's own conceit.

Envious of the attention centered on the destruction of the Federal Building in Oklahoma City, and anxious to reclaim the limelight for himself, he threatened to bring down a commercial airliner unless the *Washington Post* and the *New York Times* published a 35,000-word screed of bright but twisted self-justification.

The publishers of both papers, after consulting with the U.S. Attorney General and the director of the FBI, decided to gamble on the chance of someone recognizing the angry sentiments and the arrogant assumptions of the manifesto.

Someone did. The bomber's younger brother, who worked at a center for runaway children in New York state, remembered Ted writing something similar to local newspapers when he was a student at Harvard. A smart student already adept at hiding in plain sight, making no close friends, casting no shadow, making no waves, leaving no fingerprints. But, ultimately, betraying himself by his own prose style.

The accused Unabomber is a classic paranoiac: a bright, white, loner, quiet and unassuming, but consumed by self-importance and made destructive by self-imposed insignificance. A very small genetic Us amongst the overwhelming, unrelated Them. Someone who finds solace in isolation, who needs to be alone, but has an equally compelling need for recognition. And who conflates this contradictory situation by assuming plural pronouns for himself and dismissing all those with whom he disagrees, virtually everyone else in the world, as a singular "techno-nerd" or a "moron."

Combine this sort of passion with a culture in which secrecy prevails and in which guns are readily available, or with one that stands still in thrall to a long martial arts tradition, and you have a potent recipe for disaster. One we need to take seriously, because we are all involved in a struggle between the old genes and our new morality. Between an immoral, mindless genetic government that is truly involved in sinister plots to transform anything remotely alien into an Outsider that needs to be bashed, and a newly discovered decency that is not quite so quick to discriminate between Us and Them.

The rule, as George Williams suggests, is: "Beware of manipulation by selfish individuals, or selfish institutions, or your own selfish genes."[420] Beware particularly of any intuition which seems to follow a self-evident moral principle such as "natural justice." Beware of those who appeal to biological forces inside you, such as nepotism, in support of their "patriotic" aims. Watch out for any movement calling itself a "brotherhood" or a "sisterhood"—these tend merely to be institutional forms of the old genetic enemy pretending to be one of us. Life is rife with such unconscious traitors in our midst, but we need to confront them if we wish to become able to espouse ideals, or to pursue moral values, over which natural selection has little or no control. It can be done. We can still rise above human nature, even though sometimes it feels disloyal to do so.

Annie Dillard agonizes: "I came from the world, I crawled out of a sea of amino acids, and now I must whirl around and shake my fist at that sea and cry shame!"[109] I'm afraid so, because in a limited, but very real, sense we are moral creatures in an immoral world. There was no other way it could have been. Natural selection was necessary, and is necessarily blind. A blind man married to a beautiful woman; and we are the products of that union. The unplanned children of a world that made a pact with the devil, simply because it had to. There were no better offers on the table. So here we are now, free and seeing, faced with a blind parent and forced, not just to shake our fists, but to outwit him at every turn if we want to survive.

The new insights of evolutionary biology help us to understand the nature of the dilemma, but don't provide any easy answers. Those we will have to discover for ourselves, once we get over the astonishment of being the first creatures to see their creator clearly. This business of having the knowledge of good and evil isn't easy. The serpent lied about it. But then, being "more subtile than any beast" in the genetic field, he would, wouldn't he? That particular deceit seems, however, to have backfired on him because it was that lie, not the truth, that set us free. And we are on the brink of a freedom that has come, not just unexpectedly, but very quickly in evolutionary terms. It is something for which we are in some ways still ill-prepared and we are going to need a lot more than fig leaves to protect us out here in the cold, east of Eden. Especially now when it is becoming obvious that the enemy we face is not a chill wind or a fire-breathing demon, but something already inside us, pulling us apart.

That understanding makes evil more of a psychological problem than a theological one. We are faced with the difficult and painful acceptance of a personal Hyde to go with each of our more familiar Jekylls, and with the recognition of it as the Other that Jung called our individual shadow. And if we add genetic insights to Jung's construct, what stands revealed is a shadow that is inevitably selfish, angry, jealous, lustful, greedy, suicidal and murderous; as well as infantile, emotional, vital, neurotic and creative in an unfocused sort of way. All of which makes this Other an essential foil for the polite and generous persona which carries all the moral baggage that our parents, teachers and friends provide for us, if we are lucky, in the process of becoming encultured.

Everything Jungians believe about the shadow—that it is dangerous, disorderly, fugitive, distasteful, sensual, stupid and lacking in spirituality—is true also of the genes. Every time we laugh when someone slips on a banana peel, it is the shadow showing. Every time we take pleasure in the pain of a rival, it is a genetic pleasure. Each time we become impulsive, display exaggerated feelings about others, feel humiliated, find excessive fault, display unreasonable anger, or behave "as if we are not ourselves," we are seeing the genetic shadow in action.[439]

It can be frightening, even shocking, to come face to face with our dark side in these ways; but it is necessary. We need to leaven the single-mindedness of our unconscious attitudes with conscious flexibility. We have to confirm the dark shadow and confront its genetic troops with the light of day. We must acknowledge the beast and harness our new intellect to its old raw power. Together, we could be formidable. Apart, we are condemned to unending cycles of personal destructiveness and collective violence—unchecked evil in action.

Thomas Henry Huxley would be happy to know that the battle has been joined, but it is still at a very early stage. A peculiarly awkward and dangerous stage, because it remains difficult for us to recognize the enemy with any clarity. It is a little like trying to look through a mirror, past the image of oneself and into the eyes and mind of another. An Other who stares back at us without feeling, as it does through the eyes of Eric Smith and Robert Thompson.

I don't think it is productive or necessary to equate that look, our capacity for evil, with anything demonic. For a long time demons have served as scapegoats and repositories for all sorts of unacceptable and threatening impulses and actions; but most, if not all, of these seem now to be more sensibly equated with intrusions from our own unconscious processes. The recent reappearance of the devil as a force in our lives seems to be the result of a lack of any more psychologically accurate, integrating and meaningful myth. The convenient old timeworn symbol of the devil slides easily into such a vacuum. We need something to counter the dismay we feel at the violence and evil which appear to be epidemic in these well-newsed days. But the sudden resurgence of the ancient name has caught most of us culturally unprepared and it appears, in some quarters, to have led to the sort of morbid fascination that results in an apparent proliferation of Satanic cults.[439]

I have, as part of my normal curiosity about some of the strange things we get up to, looked closely at some of these cults and their rites, and have to say that most represent nothing more than desperate and misdirected attempts to find some personal significance; to create some sense of belonging, in otherwise very dreary lives. They appear, I believe, largely as an artifact of a western religious and ethical tradition that insists incorrectly on condemning everything that seems antisocial, and consigning it automatically to the realm of evil.[108] The dualism of good is good and evil is evil, is unhelpful, possibly even dangerous, encouraging violent eruptions of anger and destruction by participating in the process of evil, giving that which is negative unnecessary credibility. All of which tears us even further apart, when what we most need now is an awareness that true health lies in integrating good and evil, nurture and nature, the moral with that which just happens to be genetically expedient.

The mistake lies mainly in seeing our darker side as organized, something that requires a face and a name. The genes are not smart enough to get organized in that way. They have no plan, no long-term goal. They need fighting, but ought not to be misjudged, given any more credit than they deserve. Or, indeed, any more blame. Mr. Hyde behaves in an evil fashion, but it is Dr. Jekyll who is in fact the source of all evil in their relationship. Jekyll's "goodness" is a damaging distortion of what should be. He is "sinful" in the original sense of the word, which once described something that was not necessarily bad, but which "missed the mark." Something which fails to find its proper level, that position in which it is "just right." The Goldilocks solution. Things are therefore good, biologically and morally appropriate, when they are simply what they are meant to be—no more and no less.

Any deviation from this ideal is literally monstrous. Hence Mr. Hyde, Frankenstein's creation, and all the other creatures with which we choose to frighten ourselves. These monsters are essential to our health. They dramatize the evil possibilities in us in ways that remind us of the need to choose between good and evil at every juncture. It is no accident either that we find ourselves feeling somewhat ambivalent toward them, having more than a sneaking admiration for Hannibal the Cannibal, rooting for King Kong, Godzilla or the Creature from the Black Lagoon in their struggles against the mindless machineries of a state which is compelled to destroy anything it does not understand. "Monsters exist," said Primo Levi, "but they are too few in number to be truly dangerous. More dangerous are the common men, the functionaries ready to believe and to act without asking questions . . . "[241]

Genes are the enemy only when their selfishness interferes with greater happiness. Evil is undesirable only when it prevents us from being as good as we can be. Trying to rid the world of genetic or evil influences is impossible and as pointless as trying to keep weeds out of a garden. A "weed," anyway, is little more than a flower that happens to be growing in the wrong place. In their proper place, such as the fields of Morocco or the desert margins of Namaqualand, they flourish in such colourful profusion after the first spring rains that they carpet earth in the kind of glory few gardens ever achieve.

Gardeners choose, as we all must, from a catalog of options; letting these selections be acted upon in turn by weather, nature and a little judicious pruning. This is how life goes on. We give a little, take a little, try to find the best accommodation we can between mindless biological pressures and thoughtful cultural restraints. Most ordinary, decent people succeed in reaching some kind of compromise, but the fact remains that it is exactly such people, those like you and me, who have in this century alone been beguiled into the massacre of well over 100 million other ordinary, decent people. And this, in the final analysis, is far more terrifying, and much more difficult to understand, than all the muggings, murders, robberies and rapes that fatten criminal statistics and provoke such civic angst. We can distance ourselves from inner-city violence by moving to gated and guarded developments in the suburbs, but there is no way we can hide from ourselves. Or escape the fact that: "Ordinary sorts of people, using ordinary behavior, have contributed to extraordinary evil, creating major horrors."[223]

The most successful executioner ever, the man who was directly responsible for putting over 2 million people to death, was a mild-mannered "very ordinary little man" called Rudolf Hoess. He was a Catholic, the product of a devoutly religious family who taught him obedience to authority; a boy who loved nature, took up farming and was happily married with five children. He was a perfect bureaucrat who, in 1940, was given the task of designing and running Auschwitz extermination camp by an administration that believed this job required "the same attitude of impersonal rationality" as was involved in the successful running of any big farm or large company.[327]

Hoess complained later that his resources were limited, good staff hard to find, and that Berlin never gave him sufficiently explicit instructions. So he was forced to improvise. Which he did, willingly and innovatively, organizing mass murder on the grandest scale ever; and taking a bureaucrat's pride in a job well done. He was a career man in the business of genocide, exercising inflexible harshness, intent only in maximizing the product—which consisted almost entirely of dead Jews and Gypsies. He carried out his monstrous assignment with pride in being able to contribute to Hitler's noble cause of a purified and victorious Germany. And he maintained, all through this process, a sheltered home in a pretty garden near the camp, from which his children saw him off to work each morning, and welcomed their hard-working father home each evening, to play with them, help with their homework and sleep with their mother, who believed their quiet family life was a bit of paradise.[189]

Sociologist Fred Katz, himself a Holocaust survivor, says: "Most of us would prefer to believe that a man like Hoess was lying when he claimed to be sensitive to human suffering. We would assert that Hoess was not only *not* a sensitive human being but, on the contrary, he was among the coldest, the most vicious, and most evil members of the human species. He was a monster in human disguise." But, Katz admits, though this may be a plausible answer, it explains nothing. The far more chilling possibility is that Hoess was "a person operating from within the range of ordinary human behavior."[223]

Hoess was able to find satisfaction in both his work at Auschwitz and his quiet home life, though the chimneys of the crematoria were clearly visible from his bedroom window. He managed to contrive a separation between the two that was purely mental, living in one reality that was bestial and vile, and another that was nurturant and humane, commuting daily between the two for five productive years. And he achieved this apparently impossible feat by packaging everything that happened in the camp in a way that emphasized nationally approved goals and Nazi ideology. What they were doing, he convinced himself, was not immoral or deviant, but part of a grand plan that was designed to defeat his country's mortal enemies. He saw himself as a good soldier and brought a traditional German work ethic to his camp by putting up over the entrance that infamous sign that read *Arbeit Macht Frei*—"work makes one free"; and by instilling a profane kind of pride in the fact that the killings at Auschwitz were the best that modern technology could devise. Auschwitz met its deadlines.

The sheer scale of the evil industry in which Auschwitz was involved inevitably warped those taking part. A "culture of cruelty" grew up among the guards, who competed with each other in acts of savagery, deliberately vying for supremacy, for selection as "employee of the month," in a variety of grotesque specialities. One killed with the edge of his hand, another by trampling, a third by smashing children's heads against a wall. They gloried in their ghoulish reputations, proud to be known as "death on wheels," "the howling death" or "death in kid gloves." And they boasted of their relative brutality, making it quite clear in the process that they knew that what they were doing was evil, by comments such as: "In me, you will get to know the devil."[284] Hoess worried about such aberrations only when they threatened to slow down the production lines. Good help was indeed hard to find.

Those responsible for Dachau and Sachsenhausen, Treblinka and Belzec knew that what they were doing was unusual, but justified it in the larger interest of the war in which they were involved. William Calley and his platoon of United States Marines, who went into the Vietnamese village of My Lai on March 16, 1968, were in a somewhat different position. They were angry and frustrated by having to fight an enemy they could never see; they were able to measure their success only in contrived "body counts"; and they had been led to expect a full-fledged battle that day. So when, after a heavy barrage, they entered My Lai and found nothing there but unarmed old people, women and children, they went crazy and slaughtered the lot.[182]

Accidents happen in the confusion of war, but this was no incident in the heat of battle. This was a massacre. A three-hour orgy of killing in which innocent villagers were rounded up, shot and bayoneted to death by young men who enjoyed what they were doing. Who laughed and joked, firing grenade launchers indiscriminately at bunkers full of women or at water buffalo, saying: "Hey, I got me another one," and "Chalk one up for me!" It was a festive occasion, on which decent, well-trained human beings took pleasure in doing something against their entire upbringing. They found joy in their license to kill, in officially sanctioned cruelty. And, with very few exceptions, had no awareness whatsoever of behaving in an evil way. Lieutenant Calley, when ordered back to the United States months later, thought he was going back to receive a medal.[331]

These evidences of our malleability, of our ability to be flexible enough to justify just about anything we do, have been verified as universal human traits by American psychologist Stanley Milgram in a series of classic experiments. He set up a laboratory situation in which subjects thought they were teachers in a learning experiment. If the "learner" made a mistake, the "teacher's" task was to adminster an electric shock. And if the learner continued to make mistakes, the teacher was instructed to increase the power of the shock until the learner was in considerable pain. In fact there were no shocks at all, the whole experiment was simulated, sometimes even with actors playing the learner roles and giving every evidence of being severely hurt by the shocking punishments. The subjects knew nothing of this, however, and were told by those running the experiments only that they were to obey all instructions without question and that, by doing so, they would be making an important contribution to science.[272]

A few subjects refused to go through with the procedure, but most proved surprisingly willing to inflict severe pain on complete strangers, "for the sake of science." And everywhere the experiment has been repeated—in Italy, Germany, Australia and South Africa—the results have been the same. We tend, as a species, to be remarkably obedient to authority and perfectly capable, depending on the context of the situation, of doing horrible things to others. And, as My Lai has shown, we can even be persuaded to kill in cold blood, overriding and excluding all our normal inhibitions in particular situations where other values apparently hold sway.

One doesn't have to be a committed Nazi or a psychopath to do evil things. As Aleksandr Solzhenitsyn says: "If only there were evil people somewhere, insidiously committing evil deeds, and it were necessary only to separate them from the rest of us and destroy them. But the line dividing good and evil cuts through the heart of every human being. And who is willing to destroy a piece of his own heart?"[360]

That probably isn't necessary. But it is important to appreciate that we mystify evil at our peril. It is easy to dismiss the boys who killed James Bulger as "evil" and lock them up forever, but that doesn't stop such things from happening again. If there is one thing we should have learned from evil acts, it is that they tend not to be committed by extraordinary villains or by unimaginable devils or aliens, but by perfectly ordinary people. At Adolf Eichmann's trial in Jerusalem in 1960 the prosecution hoped to use him to personify the horrors of the Holocaust and made every effort to bring out the evil in the man, but he turned out to be a disappointingly ordinary person. Hannah Arendt subtitled her book about the trial *The Banality of Evil,* saying: "The trouble with Eichmann was precisely that so many were like him, and that the many were neither perverted, nor sadistic, that they were, and still are, terribly and terrifyingly normal."[14]

We all are. But we are also part of a process that recognizes what is "ugly" and doesn't quite fit. We know instinctively when something is not "just right." And being part of a rich and diverse system whose machinery is geared toward an equilibrium that has a beauty all its own, we have a finely tuned sense of what is "right" and what is "wrong."

We happen also to be the products of natural selection, of genes that are essentially and necessarily selfish, with a rigid set of rules that maximize a very narrow kind of short-term advantage. It is a system that works wonderfully well, on its own limited terms. It has produced everything in us that is both "good" and "evil," but it is completely amoral, devoid of empathy and long-term concern.

It is up to us to provide these moral qualities, to give life on Earth a conscience. We are this world's first ethical animals, at the mercy still of our biology, but capable also of rising above it. Intelligence helps. It succeeds, at the very least, in widening our view. It gives us access to a whole range of new horizons, new possibilities. It allows us some control over the direction in which our moods may take us. It is, in philosopher Mary Midgley's view, nature's way of dealing with "emotional tunnel vision."[271]

Wide horizons, however, also open the possibility of choice and conflict, something with which most other species don't have to deal. We do. But the intelligence which created the problem also goes some way toward solving it, by the use of imagination and forethought and forward planning. These all make a considerable difference, but there are always going to be situations in which even they fall short; and that is where morality comes in. A system of priorities which helps to indicate which motives *ought* to prevail.

This is how Darwin saw morality, as mankind's own necessary remedy for internal conflicts. "It is very probable," he said, "that any animal whatever, endowed with well-marked social instincts, would inevitably acquire a moral sense or conscience, as soon as its intellectual powers had become as well-developed, or nearly as well-developed, as in man."[97]

That feels right, and the evolution of a moral sense, and of the necessary immoral acts which sharpen such a sense, are just as likely wherever life may be found. That is good biology, but problems arise and lie in the slippery and imperious word "ought." Darwin saw morality with Victorian eyes as a natural result of our discomfort with conflicting and often destructive motives; a dilemma capable of being resolved by an exercise of will—even if this involved a choice between greater or lesser evils. But "ought" suggests more than an assessment of relative probabilities, more than a gamble on possible profit and loss. It implies an external imperative, a road already chosen for us, leading to ready-made and totally satisfactory solutions. The best of all possible worlds. And given everything we now know about how the world works, that seems utopian and unlikely.

The best we can hope for is to exercise our freedom to choose with the advantage of as much knowledge as may be available. But we mustn't expect any help from our institutions.

Capitalism operates, almost by definition, on the basis of what is expedient. So everything in economics is wide open to question on ethical grounds. Like natural selection, it is self-seeking and so obsessed with selective advantage and the profit motive that it cannot be trusted to provide moral guidance. Socialism, on the other hand, wears a kinder face, but is so out of touch with basic biology that it doesn't work at all.

The law seems to have scruples and pays lip service to justice, but operates so often in lock-step with the prevailing social climate and at the behest of politics that its motives too are necessarily suspect. It is particularly weak in its approach to strong evil, preferring, it seems, to play to the genetic gallery by prosecuting and punishing petty crime.

Science is now so morally unconstrained that it tends to devise and to employ new techniques just because it can, without considering the consequences or asking whether it should. And it has become, in all too many areas, a slave to industry, far too often providing "experts" willing to testify to whatever truths vested interests want to hear. It seems, in this respect, to have fallen into the same trap as much of organized religion, becoming an establishment intent more on protecting its own interests than in looking at the world with curiosity and honesty.

All we have left, it appears, is ourselves. Our divided selves, fighting individual battles on a very wide front, using what in the end may be our best weapons. Reviving neglected aspects of our own biological inheritance, reänimating the world and rediscovering long-dormant faculties, using ourselves as the ultimate instruments of knowing.

This is something that comes naturally to us, and we can be very good at it indeed. Perhaps even good enough to hold the genetic tide, and our own dark nature, at bay long enough to give evolution the nudge it needs in the right direction.

The choice is ours. It is the capacity to choose that makes us special, giving us the ability to select a course for nature, instead of just submitting to the course of natural selection.

BIBLIOGRAPHY

1. Adele, L.G. & Gilchrist, S., "Homosexual rape and sexual selection," *Science*, 197: 81, 1977.
2. Adler, M.J., *Desires Right and Wrong*, Macmillan, New York, 1991.
3. Alexander, R.D. & Tinkle, D.W., *Natural Selection and Social Behavior*, Chiron, New York, 1980.
4. Altman, J., *Baboon Mothers and Infants*, Harvard University, 1980.
5. Alvarez, F. et al., "Experimental brood parasitism . . . ," *Animal Behaviour*, 24: 907, 1976.
6. Amis, K. & Conquest, R. (eds), *Spectrum V*, Harcourt, Brace, New York, 1966.
7. Amis, M., "Blown away," *New Yorker*, 30 May 1994.
8. Anderson, E., "The code of the streets," *Atlantic Monthly*, May 1994.
9. Andersson, M., "Female choice . . . ," *Nature*, 299: 818, 1982.
10. Angier, N., "The debilitating malady . . . ," *New York Times*, 17 July 1994.
11. Appleyard, B., "The worst murder. . . ," *Independent*, 25 November 1993.
12. Apsche, J.A., *Probing the Mind of a Serial Killer*, International, Philadelphia, 1993.
13. Archimedes, *The Works of Archimedes*, Dover, New York, 1953.
14. Arendt, H., *Eichmann in Jerusalem*, Penguin, Harmondsworth, 1964.
15. Aristotle, *The Nicomachean Ethics*, Oxford University, 1925.
16. Associated Press report, 1 July 1994.
17. Audubon, J.J., *Ornithological Biography*, London, 1839.
18. Aureli, F. et al., "Kin-oriented redirection . . . ," *Animal Behaviour*, 44: 283, 1992.
19. Austin, A.L., Letter to Charles Darwin, 6 November 1877.
20. Axelrod, R., "The emergence of cooperation . . . ," *American Political Science Review*, 75: 306, 1981.
21. Axelrod, R., *The Evolution of Cooperation*, Basic, New York, 1984.
22. Baker, R.R. & Bellis, M.A., "Number of sperm in human ejaculates," *Animal Behaviour*, 37: 867, 1989.
23. Baker, R.R. & Bellis, M.A., "Human sperm competition," *Human Behavior and Evolution Society*, New Mexico, July 1992.
24. Baker, R.R., *Human Navigation*, Hodder & Stoughton, London, 1981.
25. Barash, D.P., "Sociobiology of rape . . . ," *Science*, 197: 788, 1977.
26. Barash, D.P., *Sociobiology*, Harper & Row, New York, 1980.
27. Barkow, J.H. et al. (eds), *The Adapted Mind*, Oxford University, 1992.

28. Barrett, L. & Dunbar, R., "Not now dear . . . ," *New Scientist,* 9 April 1994.
29. Barsley, M., *The Left-handed Book,* Souvenir, London, 1966.
30. Bateson, P.G., "Is aggression instinctive?," in Groebel & Hinde (Ref. 162).
31. Bateson, P.G. & Hinde, R.A., *Growing Points in Ethology,* Cambridge University, 1976.
32. Beattie, J.M., "The pattern of crime in England," *Past and Present,* 62: 47, 1974.
33. Bedell, G., "The death of innocence," *Independent on Sunday,* 28 November 1993.
34. Berkowitz, L., *Roots of Aggression,* McGraw Hill, New York, 1964.
35. Bernds, W.P. & Barash, D.P., "Early termination . . . ," in Chagnon & Irons (eds) (Ref. 66).
36. Betzig, L.L., *Despotism and Differential Reproduction,* Aldine, New York, 1986.
37. Betzig, L.L., "Roman polygyny," *Ethology and Sociobiology,* 13: 309, 1992.
38. Betzig, L.L., "Medieval monogamy," in Mithen & Maschner (Ref. 273).
39. Birkhead, T.R. & Moller, A.P., *Sperm Competition in Birds,* Academic, London, 1992.
40. Blashfield, J.F., *Why They Killed,* Warner, New York, 1990.
41. Bleier, R., *Feminist Approaches to Science,* Pergamon, New York, 1986.
42. Blick, J.P., "Genocidal warfare . . . ," *Man,* 23: 654, 1988.
43. Bly, R., *A Little Book on the Human Shadow,* HarperCollins, New York, 1988.
44. Boehm, C., "Segmentary warfare . . . ," in Harcourt & de Waal (Ref. 174).
45. Bohannen, P. (ed), *Law and Warfare,* New York, 1966.
46. Brownmiller, S., *Against Our Will,* New York, 1975.
47. Brownmiller, S., "Making female bodies the battlefield," in Stiglmayer (Ref. 367).
48. Brunner, H.G. et al., "Abnormal behavior . . . ," *Science,* 262: 578, 1994.
49. Bunyan, N., "Boys guilty . . . ," *Daily Telegraph,* 25 November 1993.
50. Burch, T.K., "Family history survey," *Statistics Canada,* Ottawa, 1985.
51. Burgess, A., *A Clockwork Orange,* Heinemann, London, 1962.
52. Burgess, A., *Little Wilson and Big God,* Heinemann, London, 1978.
53. Burkert, W., *Homo Necans,* University of California, 1983.
54. Burl, A., *The Stonehenge People,* Dent, London, 1987.
55. Burnstein, M.H. et al. (eds), *Psychological Development from Infancy,* Erlbaum, New Jersey, 1979.

56. Buskirk, R. et al., "Sexual cannibalism . . . ," *American Naturalist,* 123: 612, 1984.
57. Bygott, D., "Cannibalism among wild chimpanzees," *Nature,* 238: 410, 1972.
58. Byrne, R. & Whiten, A., "Tactical deception . . . ," *Animal Behaviour,* 33: 669, 1985.
59. Byrne, R. & Whiten, A., *Machiavellian Intelligence,* Oxford Scientific, 1988.
60. Byron, Lord, *The Destruction of Sennacherib,* London, 1818.
61. Caughley, G., "The elephant problem . . . ," *East African Wildlife Journal,* 14: 265, 1976.
62. CBS Television, *48 Hours Special Report,* 17 August 1994.
63. Chagnon, N.A., *Yanomamö: The Fierce People,* Holt Rinehart, New York, 1977.
64. Chagnon, N.A., "Life histories . . . ," *Science,* 239: 935, 1988.
65. Chagnon, N.A., "Reproductive and somatic conflicts . . . ," in Haas (Ref. 164).
66. Chagnon, N.A. & Irons, W., *Evolutionary Biology and Social Behaviour,* Duxbury, Massachusetts, 1979.
67. Charlson, R.J. et al., "Ocean phytoplankton . . . ," *Nature,* 326: 655, 1987.
68. Cheyney, D.L. & Seafarth, R.M., "Vervet monkey alarm calls . . . ," *Behaviour,* 94: 150, 1985.
69. Christiansen, K.O., "Criminality among twins," in Mednick & Christiansen (Ref. 269).
70. Churchill, W.S., *The Second World War,* London, 1954.
71. Clark, R. W., *The Survival of Charles Darwin,* Arrow, New York, 1986.
72. Clendinnen, I., *Aztecs,* Cambridge University, 1991.
73. Clutton-Brock, T.H. et al., "The logical stag," *Animal Behavior,* 27: 211, 1979.
74. Coale, A.J., "The history of the human population," *Scientific American,* 23: 40, 1974.
75. Cohn, N., *Europe's Inner Demons,* Sussex University, 1975.
76. Coleridge, S.T., *Notes on the Tragedies of Shakespeare,* London, 1826.
77. Collett, P. (ed), *Social Rules and Social Behaviour,* Oxford University, 1976.
78. Contini, P., "The evolution of blood money . . . ," *Journal of African Law,* 15: 77, 1971.
79. Cook, T.A., *The Curves of Life,* Dover, New York, 1979.
80. Corballis, M.C., "Laterality and human evolution," *Psychological Review,* 96: 492, 1989.
81. Corballis, M.C., *The Lopsided Ape,* Oxford University, 1991.

82. Coren, S., *The Left-handed Syndrome,* Vintage, New York, 1993.
83. Cox, C.R. & Le Boeuf, B.J., "Female imitation . . . ," *American Naturalist,* 111: 317, 1977.
84. Cuff, W.R., "Behavioural aspects of cannibalism . . . ," *Canadian Journal of Zoology,* 58: 1504, 1980.
85. Dahmer, L., *A Father's Story,* Morrow, New York, 1994.
86. *Daily Mail,* London, 19 March 1973.
87. Dalgaard, O.S. & Kringlen, E., "A Norwegian twin study . . . ," *British Journal of Criminal Psychology,* 16: 213, 1976.
88. Daly, M., "The cost of mating," *American Naturalist,* 112: 771, 1978.
89. Daly, M. & Wilson, M., "Homicide and kinship," *American Anthropologist,* 84: 372, 1982.
90. Daly, M. & Wilson, M., "A sociobiological analysis . . . ," in Hausfater & Hrdy (Ref. 180).
91. Daly, M. & Wilson, M., "Child abuse . . . ," *Ethology and Sociobiology,* 6: 197, 1985.
92. Daly, M. & Wilson, M., "Evolutionary social psychology . . . ," *Science,* 242: 519, 1988.
93. Daly, M. & Wilson, M., *Sex, Evolution and Behavior,* Willard Grand, Boston, 1983.
94. Daly, M. & Wilson, M., *Homicide,* Aldine, New York, 1988.
95. Damasio, H. et al., "The return of Phineas Gage . . . ," *Science,* 264: 1102, 1994.
96. Darwin, C., *The Voyage of the Beagle,* Dent, London, 1906.
97. Darwin, C., *The Descent of Man,* John Murray, London, 1871.
98. Davies, N., *Human Sacrifice,* Morrow, New York, 1981.
99. Davies, P.C.V., *The Accidental Universe,* Cambridge University, 1982.
100. Davis, M. et al., "Extinction of species . . . ," *Nature,* 308: 715, 1984.
101. Dawkins, R., *The Selfish Gene,* Oxford University, 1976.
102. Dawkins, R., *The Extended Phenotype,* Oxford University, 1982.
103. Deacon, E.L., *Boundary Layer Meteorology,* 17: 517, 1979.
104. Degler, C.N., *In Search of Human Nature,* Oxford University, 1991.
105. Den Boer, P.J. & Gradwell, G.R. (eds), *Dynamics of Populations,* Wageningen, Holland, 1971.
106. Dentan, R.K., *The Semai,* Holt Rinehart, New York, 1968.
107. Devereux, G., "Cannibalistic impulses . . . ," *Psychoanalytic Forum,* 1: 114, 1966.
108. Diamond, S.A., "Redeeming our devils . . . ," in Zweig & Abrams (Ref. 439).
109. Dillard, A., *Pilgrim at Tinker Creek,* Harpers, New York, 1974.
110. Dirac, P.A.M., "The cosmological constants," *Nature,* 139: 323, 1937.

111. Dostoyevsky, F., *The Brothers Karamazov,* 1879.
112. Downhower, J.F. & Brown, C., "The timing of reproduction . . . ," in Alexander & Tinkle (Ref. 3).
113. Eggan, D., "Hopi adjustment . . . ," *American Anthropologist,* 45: 372, 1943.
114. Ehrlich, A.H., "The human population," *American Zoologist,* 25: 395, 1985.
115. Ehrlich, P.R., *The Machinery of Nature,* Touchstone, New York, 1986.
116. Eibl-Eibesfeldt, I., "Expressive movements . . . ," in Hinde (Ref. 185).
117. Eibl-Eibesfeldt, I., *The Biology of Peace and War,* Thames & Hudson, London, 1979.
118. Eigen, M. & Winkler, R., *Laws of the Game,* Allen Lane, London, 1982.
119. Emlen, S.T., "Altruism, kinship and reciprocity . . . ," in Alexander & Tinkle (Ref. 3).
120. Englert, P.S., *Islands at the Center of the World,* Scribners, New York, 1990.
121. Estes, J.A. et al., "Activity and prey selection . . . ," *American Naturalist,* 120: 242, 1982.
122. Euripedes, *Medea* (trans. Alistair Elliott, 1993).
123. Evans-Pritchard, E.E., *Witches, Oracles and Magic among the Azande,* Clarendon, London, 1937.
124. Ewart, A., *The World's Wickedest Men,* Odhams, London, 1972.
125. Falger, V.S.E., "Cooperation in conflict," in Harcourt & de Waal (Ref. 174).
126. Firth, R., "Reason and unreason . . . ," in Marwick (Ref. 265).
127. Forleo, R. & Pasini, W., *Medical Sexology,* PSG, Littleton, 1983.
128. Foster, J., "Fatal meeting . . . ," *Independent,* 25 November 1993.
129. Fossey, D., "Living with mountain gorillas," in Marler (Ref. 260).
130. Fox, R., "The inherent rules of violence," in Collett (Ref. 77).
131. Frank, L., "When hyenas kill their own," *New Scientist,* 5 March 1994.
132. Frazer, J.G., *The Dying God,* Macmillan, London, 1911.
133. Frazer, J.G., *The Scapegoat,* Macmillan, London, 1914.
134. Freud, S., "Why war," *Collected Papers,* 5: 273, 1959.
135. Friedmann, L., *The Law of War,* Random House, New York, 1972.
136. Fromm, E., *The Anatomy of Human Destructiveness,* Jonathan Cape, London, 1974.
137. Galton, F., "Composite portraits," *Nature,* 18: 97, 1878.
138. Gelles, R.J. (ed), *Family Violence,* Sage, Beverly Hills, 1979.
139. Gelles, R.J. & Strauss, M.A., "Family experience . . . ," in Gelles (Ref. 138).

140. Gibbs, N., cover story, *Time* magazine, 16 May 1994.
141. Gibbs, N., "Death and deceit," *Time* magazine, 14 November 1994.
142. Gifford, E., *The Evil Eye,* Macmillan, New York, 1958.
143. Gilliard, E.T., *Living Birds of the World,* Hamish Hamilton, London, 1958.
144. Giovanni, J. Di, "Sinister forces . . . ," *Sunday Times,* 31 October 1993.
145. Girard, R., *Violence and the Sacred,* John Hopkins University, 1977.
146. Given, J.B., *Society and Homicide in Thirteenth Century England,* Stanford University, 1977.
147. Glance, N.S. & Huberman, B.A., "The dynamics of social dilemmas," *Scientific American,* March 1994.
148. Glover, S., "A story devoid of mercy or hope," *Evening Standard,* 25 November 1993.
149. Golding, W., *Lord of the Flies,* Coward McCann, New York, 1962.
150. Goldschmidt, W.R., *Sebei Law,* University of California, 1967.
151. Goodall, J., *In the Shadow of Man,* Collins, London, 1971.
152. Goodall, J., *The Chimpanzees of Gombe,* Harvard University, 1986.
153. Goodall, J. et al., "Intercommunity interaction . . . ," in Hamburg & McCown (Ref. 166).
154. Goode, W., "Violence among intimates," in Mulvihill et al. (Ref. 280).
155. Goslin, D.A. (ed), *Handbook of Socialisation Theory . . . ,* McNally, Chicago, 1969.
156. Gould, J.L. & C.G., *Sexual Selection,* Freeman, New York, 1989.
157. Green, R., *Sexual Science and the Law,* Harvard University, 1993.
158. Green, R. et al., "Shock disease . . . ," *American Journal of Hygiene,* 30: 83, 1939.
159. Greenland, C. & Rosenblatt, E., "Murder followed by suicide . . . ," in Daly & Wilson (Ref. 94).
160. Gregor, T., "Uneasy peace . . . ," in Haas (Ref. 164).
161. Gribbin, J., *In the Beginning,* Viking, London, 1993.
162. Groebel, J. & Hinde, R. (eds), *Aggression and War,* Cambridge University, 1989.
163. Grove, W.M. et al., "Heritability of antisocial behavior," *Biological Psychiatry,* 27: 1293, 1990.
164. Haas, J., *The Anthropology of War,* Cambridge University, 1990.
165. Haldane, J.B.S., "Population genetics," *New Biology,* 18: 34, 1955.
166. Hamburg, D. & McCown, E. (eds), *The Great Apes,* Benjamin Cummings, Menlo Park, 1979.
167. Hamilton, W.D., "The evolution of altruistic behaviour," *American Naturalist,* 97: 1, 1963.
168. Hamilton, W.D., "The evolution of social behaviour," *Journal of Theoretical Biology,* 7: 1, 1964.

169. Hamilton, W.D., "Selfish and spiteful behaviour . . . ," *Nature,* 228: 1218, 1970.
170. Hamilton, W.D., "Geometry for the selfish herd," *Journal of Theoretical Biology,* 31: 295, 1971.
171. Hamilton, W.D., "Sex versus non-sex versus parasite," *Oikos,* 35: 282, 1980.
172. Hamilton, W.D. & Zuk, M. "Heritable true fitness and bright birds," *Science,* 218: 384, 1982.
173. Hanson, V., *Warfare and Agriculture in Classical Greece,* New York, 1983.
174. Harcourt, A.H. & De Waal, T.B.M., *Coalitions and Alliances in Humans and Other Animals,* Oxford University, 1992.
175. Hardy, T., *The Dynasts,* London, 1903.
176. Harris, M., *Cows, Pigs, Wars and Witches,* Random House, New York, 1974.
177. Harrison, D., "Consumed in the bonfire . . . ," *Observer,* 28 November 1993.
178. Harrison, S., "The symbolic construction of aggression . . . ," *Man,* 24: 583, 1989.
179. Hauser, M.D., "Invention and social transmission . . . ," in Byrne & Whiten (Ref. 59).
180. Hausfater, G. & Hrdy, S.B. (eds), *Infanticide,* Aldine, New York, 1984.
181. Herbert, A.J., "The physiology of aggression," in Groebel & Hinde (Ref. 162).
182. Hersh, S., *My Lai 4,* Random House, New York, 1970.
183. Hick, J., *Evil and the God of Love,* Harper, London, 1985.
184. Hinde, R.A., *Bird Vocalisations,* Cambridge University, 1969.
185. Hinde, R.A. (ed), *Non-verbal Communication,* Cambridge University, 1972.
186. Hinde, R.A., "A biologist looks at anthropology," *Man,* 26: 583, 1991.
187. Hobbes, T., *Leviathan,* London, 1651.
188. Hoebel, E.A., "Song duels among the Eskimos," in Bohannen (Ref. 45).
189. Hoess, R., *Commandant of Auschwitz,* World, New York, 1959.
190. Holy Bible, Genesis 3: 7.
191. Holy Bible, Genesis 4: 7.
192. Holy Bible, I Chronicles 21: 1.
193. Holy Bible, Numbers 31: 17–18.
194. Hooley, P., "How could it happen?," *Daily Express,* 26 November 1993.
195. Hoyle, F., *The Intelligent Universe,* Michael Joseph, London, 1983.
196. Hrdy, S.B., *The Langurs of Abu,* Harvard University, 1977.

97. Hrdy, S.B., "Infanticide as a primate reproductive strategy," *American Science,* 65: 40, 1977.
98. Hrdy, S.B., "Empathy, polyandry . . . ," in Bleier (Ref. 41).
99. Huck, H.W., "Lemming migrations," in Macdonald (Ref. 251).
00. Hughes, R., *A High Wind in Jamaica,* Collins, London, 1929.
01. Humphrey, N.K., *Consciousness Regained,* Oxford University, 1984.
02. Humphrey, N.K., "The social function of intellect," in Bateson & Hinde (Ref. 31).
03. Hungerford, M.W., *Molly Bawn,* Dublin, 1878.
04. Hunt, M., *The Mugging,* Penguin, Harmondsworth, 1976.
05. Huntingford, F.A., "Animals fight, but do not make war," in Groebel & Hinde (Ref. 162).
06. Huxley, A., *Plant and Planet,* Allen Lane, London, 1974.
07. Huxley, F., *The Eye: The Seer and the Seen,* Thames & Hudson, London, 1990.
08. Huxley, T.H., *Evolution and Ethics,* Macmillan, London, 1894.
09. Hymas, C., "Bulger: chill verdict . . . ," *Sunday Times,* 28 November 1993.
10. Icli, T.G., "Blood feud in Turkey," *British Journal of Criminology,* 34: 69, 1994.
11. *Independent,* 28 November 1993.
12. Ingold, T., "An anthropologist looks at biology," *Man,* 25: 208, 1990.
13. Jones, A., *Women Who Kill,* Fawcett, New York, 1980.
14. Jung, C.G., "On the psychology of the unconscious," *Collected Works:* Volume 7, 1917.
15. Jung, C.G., *Modern Man in Search of a Soul,* Routledge, London, 1933.
16. Jung, C.G., "Answer to Job," *Collected Works:* Volume 11, 1952.
17. Jung, C.G., *Two Essays on Analytical Psychology,* Routledge, London, 1953.
18. Jung, C.G., *Memories, Dreams, Reflections,* Random House, New York, 1961.
19. Jung, C.G., *The Visions Seminars,* Spring, Zurich, 1976.
20. Junod, T. et al., "Blown away," *GQ* magazine, July 1994.
21. Kaplan, R.D., "The coming anarchy," *Atlantic Monthly,* February 1994.
22. Kass, L.R., *Toward a More Natural Science,* Free Press, New York, 1985.
23. Katz, F.E., *Ordinary People and Extraordinary Evil,* State University of New York, 1993.
24. Kawai, M., "Newly acquired precultural behaviour . . . ," *Primates,* 6: 1, 1965.

225. Keegan, J., *A History of Warfare,* Hutchinson, London, 1993.
226. Keiser, R.L., *The Vice Lords,* Holt Rinehart, New York, 1969.
227. Kenny, M.G., "Mirror in the forest," *Africa,* 51: 477, 1981.
228. Kierkegaard, S., *Fear and Trembling,* Penguin, Harmondsworth, 1985.
229. Kilgore, B.M., "Restoring fire's natural role . . . ," *Western Wildlands,* 11: 2, 1984.
230. Kluger, H.V., *Satan in the Old Testament,* Northwestern University, Evanston, 1967.
231. Krebs, C.J., *The Message of Ecology,* HarperCollins, New York, 1988.
232. Krige, E.J. & J.D., *The Realm of the Rain Queen,* Oxford University, 1943.
233. Kuhl, P. et al., "Linguistic experience . . . ," *Science,* 255: 606, 1974.
234. Kullenberg, B., "Studies in *Ophrys* pollination," *Zooligische Bidrag,* 34: 1, 1961.
235. Kummer, H., "Analogs of morality," in Stent (Ref. 363).
236. Langlois, J.H. et al., "Infant preferences for attractive faces," *Developmental Psychology,* 23: 363, 1987.
237. Langlois, J.H., "Attractive faces are only average," *Psychological Science,* 1: 115, 1990.
238. Le Boeuf, B.J., "Male-male competition . . . ," *American Zoologist,* 14: 163, 1974.
239. Le Boeuf, B.J. & Reither, J., "The cost of living in a seal harem," *Mammalia,* 41: 167, 1977.
240. Lee, R., *The !Kung San,* Cambridge University, 1979.
241. Levi, P., *Survival in Auschwitz,* Summit, New York, 1986.
242. Lienhardt, G., "Some notions of witchcraft among the Dinka," *Africa,* 21: 303, 1951.
243. Lockard, J.S. (ed), *The Evolution of Human Social Behaviour,* Elsevier, New York, 1980.
244. Loizos, P., "Intercommunal killing on Cyprus," *Man,* 23: 639, 1988.
245. Lorenz, K., *On Aggression,* Methuen, London, 1966.
246. Low, B.S., "Marriage systems and pathogen stress . . . ," *American Zoologist,* 30: 325, 1990.
247. Low, B.S. et al., "Human hips, breasts and buttocks . . . ," *Ethology and Sociobiology,* 8: 249, 1987.
248. Maccoby, E.E. & Jacklin, C.N., *The Psychology of Sex Differences,* Stanford University Press, 1974.
249. Maccoby, H., *The Sacred Executioner,* Thames & Hudson, London, 1982.
250. McCormick, J., "Why parents kill," *Newsweek,* 14 November 1994.
251. MacDonald, D., *The Encyclopaedia of Mammals,* Allen & Unwin, London, 1984.

52. MacFarlane, A., *Witchcraft in Tudor and Stuart England,* Routledge, London, 1970.
53. McManus, I.C. & Humphrey, N.K., "Turning the left cheek," *Nature,* 243: 271, 1973.
54. McWhirter, N. & Greenberg, S., *Guinness Book of Records,* Guinness Superlatives, London, 1979.
55. Machiavelli, N., *The Prince,* Florence, 1515.
56. Maier-Katkin, D., "Mothers who kill," *Miami Herald,* 20 November 1994.
57. Mailer, N., *The Executioner's Song,* Little, Brown, New York, 1979.
58. Manning, A., "The genetic bases of aggression," in Groebel & Hinde (Ref. 162).
59. Marcy, W.L., Speech to US Senate, January 1832.
60. Marler, P.R. (ed), *The Marvels of Animal Behavior,* National Geographic, Washington, 1972.
61. Marsden, W., *The Lemming Year,* Chatto & Windus, 1964.
62. Marsh, P., *Aggro,* Dent, London, 1978.
63. Marshall, A.J., *Bowerbirds,* Oxford University, 1954.
64. Marshall, S.L.A., *Men Against Time,* Morrow, New York, 1947.
65. Marwick, M. (ed), *Witchcraft and Sorcery,* Penguin, Harmondsworth, 1972.
66. Masters, B., *Killing For Company,* Jonathan Cape, London, 1985.
67. Mayer, P., "Witches," in Marwick (Ref. 265).
68. Maynard Smith, J., *Evolution and the Theory of Games,* Cambridge University, 1982.
69. Mednick, S.A. & Christiansen, K.O., *Biological Bases of Criminal Behavior,* Gardner, New York, 1977.
70. Mestel, R., "What triggers the violence within?," *New Scientist,* 29 February 1994.
71. Midgley, M., *Wickedness,* Routledge, London, 1984.
72. Milgram, S., *Obedience to Authority,* Harper & Row, New York, 1974.
73. Mithen, S. & Maschner, H. (eds), *Darwinian Approaches to the Past,* Plenum, New York, 1992.
74. Moller, A.P., "Intruders and defenders . . . ," *Oikos,* 48: 47, 1987.
75. Montaigne, M. de, *Essays,* Paris, 1580.
76. Moore, O.K., "Divination—a new perspective," *American Anthropologist,* 59: 69, 1957.
77. Morgan, B., *Men and Discoveries in Mathematics,* John Murray, London, 1972.
78. Morris, D., *Manwatching,* Jonathan Cape, London, 1977.
79. Morris, D. et al., *Gestures,* Jonathan Cape, London, 1979.
80. Mulvihill, D.J. et al., *Crimes of Violence,* US Government Printer, Washington DC, 1969.

281. Murphy, R.F., *Headhunter's Heritage,* University of California, 1960.
282. Myers, K., "The rabbit in Australia," in Den Boer & Gradwell (Ref. 105).
283. Nash, E.P., "What's a life worth?," *New York Times* magazine, 14 August 1994.
284. Naumann, B., *Auschwitz,* Praeger, New York, 1966.
285. Newton, M., *Hunting Humans: Volumes 1&2,* Avon, New York, 1990.
286. Nightingale, B., "Yankee cinema, please go home," *New York Times,* 10 July 1994.
287. Nishida, T., "Alpha status and agonistic alliance . . . ," *Primates,* 24: 318, 1983.
288. Nitecki, M.H. & D.V., *Evolutionary Ethics,* State University of New York, 1993.
289. Norris, J., *Serial Killers,* Doubleday, New York, 1988.
290. Oates, J.C., "I had no other thrill or happiness," *New York Review,* 24 March 1994.
291. O'Connor, R.J., "Brood reduction in birds," *Animal Behaviour,* 26: 79, 1978.
292. Ogot, B. (ed), *War and Society in Africa,* Frank Cass, London, 1972.
293. Organ, J.A. & D.J., "Courtship behavior of the red salamander," *Copeia,* 217, 1968.
294. Orians, G.H. & Willson, M.F., "Interspecific territories of birds," *Ecology,* 45: 736, 1964.
295. Otterbein, K., "The evolution of Zulu warfare," in Ogot (Ref. 292).
296. Otterbein, K.F., *The Evolution of War,* HRAF, Connecticut, 1970.
297. Paradis, J. & Williams, G.C., Prefaces to *Evolution and Ethics,* Princeton University, 1989.
298. Pausanius, *Description of Greece,* London, 1890.
299. Perrett, D.I. et al., "Facial shape . . . ," *Nature,* 368: 239, 1994.
300. Persaud, R., "When pairs can turn into savages," *Daily Mail,* 25 November 1993.
301. Phillips, D.P., "The impact of mass media violence . . . ," *American Sociological Review,* 48: 560, 1983.
302. Phillips, M., "How we make demons . . . ," *Observer,* 28 November 1993.
303. Pilkington, E., "Boys guilty of Bulger murder," *Guardian,* 25 November 1993.
304. Pinker, S., *The Language Instinct,* William Morrow, New York, 1994.
305. Porac, C. & Coren, S., *Lateral Preferences and Human Behavior,* Springer, New York, 1981.
306. Porter, J.W., "*Pseudorca* stranding," *Oceans,* 10: 8, 1977.

07. Pratt, H.J., "Reproduction in the blue shark," *Fishery Bulletin,* 77: 445, 1979.
08. Raine, A., *The Psychopathology of Crime,* Academic, San Diego, 1993.
09. Raine, A., "When bad things happen . . . ," *GQ* magazine, July 1994.
10. Rampino, M.R. & Stothers, R.B., "Geological rhythms and cometary impacts," *Science,* 226: 1427, 1984.
11. Rampino, M.R. & Stothers, R.B., "Geological periodicity and the galaxy," in Smoluchowski (Ref. 359).
12. Raup, D.M. & Sepkoski, J.J., "Periodicity of extinctions . . . ," *Proceedings of the National Academy of Sciences,* 81: 801, 1984.
13. Read, P.P., "The most difficult question . . . ," *Daily Mail,* 25 November 1993.
14. Rees, M.J. et al., *Black Holes, Gravitational Waves and Cosmology,* Gordon & Breach, New York, 1974.
15. Reinhardt, J.M., *The Psychology of Strange Killers,* Thomas, Springfield, 1962.
16. Rheingold, H.L., "The social and socialising infant," in Goslin (Ref. 155).
17. Rheingold, H.L. et al., "Sharing in the second year . . . ," *Child Development,* 47: 1148, 1976.
18. Rheingold, H.L. & Hay, D.F., "Prosocial behavior . . . ," in Stent (Ref. 363).
19. Reinisch, J.M. & Karow, W.G., "Prenatal exposure to synthetic progestin . . . ," *Archives of Sexual Behavior,* 6: 257, 1977.
20. Ridley, M., "Infected with science," *New Scientist,* 1 January 1994.
21. Ridley, M., *The Red Queen,* Macmillan, London, 1993.
22. Robarchek, C.A., "Frustration, aggression . . . ," *American Ethologist,* 4: 762, 1977.
23. Robarchek, C.A. & Dentan, R.K., "Blood drunkenness . . . ," *American Anthropologist,* 89: 356, 1987.
24. Roberts, T.A. & Kennelly, J., "Variation in promiscuity . . . ," *Wilson Bulletin,* 92: 110, 1980.
25. Rosenblatt, R., "A killer in the eye," *New York Times Magazine,* 5 June 1994.
26. Routledge, W.S. & K.R., *With a Prehistoric People,* Cass, London, 1968.
27. Rubenstein, R., *After Auschwitz,* Bobbs Merrill, Indianapolis, 1975.
28. Rubin, L., *Just Friends,* Harper & Row, New York, 1985.
29. Rule, A., *The Stranger Beside Me,* Signet, New York, 1993.
30. Russell, J.B., *The Devil,* Cornell University, Ithaca, 1977.
31. Sack, J., *Lieutenant Calley,* Viking, New York, 1971.
32. Sagan, C. & Druyan, A., *Shadows of Forgotten Ancestors,* Century, London, 1992.

33. Sanford, J.A., *Jung and the Problem of Evil,* Sigo, Boston, 1987.
34. Schaffer, H.R., "Acquiring the concept of the dialogue," in Burnstein (Ref. 55).
35. Schaller, G.B., *The Serengeti Lion,* University of Chicago, 1972.
36. Schele, L. & Miller, M.E., *The Blood of Kings,* Kimbell Museum, Fort Worth, 1986.
37. Schneider, D., "Slow-motion explosion," *Whole Earth Review,* 83: 101, 1994.
38. Schnell, R.C. et al., "Biogenic ice nuclei . . . ," *15th Conference on Agriculture and Forest Meteorology,* 22, 1981.
39. Schultz, A.M., "The Arctic tundra," in Van Dyne (Ref. 394).
40. Schwartz, A.E., *The Man Who Could Not Kill Enough,* Birch Lane, New York, 1993.
41. Scott, P.D., "Parents who kill their children," *Medicine, Science and the Law,* 13: 120, 1973.
42. Seifert, R., "War and rape," in Stiglmeyer (Ref. 367).
43. Self, W., "Where did I go wrong?," *New York Times,* 13 March 1994.
44. Sereny, G., *The Case of Mary Bell,* Methuen, London, 1972.
45. Shakespeare, W., *Macbeth,* I, vii, 54.
46. Shakespeare, W., *Macbeth,* IV, i, 44.
47. Shaw, G.B., *Back to Methuselah,* Collins, London, 1921.
48. Sheldrake, R., *A New Science of Life,* Blond, London, 1981.
49. Sheldrake, R., *The Presence of the Past,* Collins, London, 1988.
50. Sheldrake, R., *The Rebirth of Nature,* Rider, London, 1990.
51. Sheldrake, R., *Seven Experiments That Could Change the World,* Fourth Estate, London, 1994.
52. Shelley, M., *Frankenstein,* Lackington, London, 1818.
53. Sherman, P.W., "Reproduction, competition and infanticide," in Alexander & Tinkle (Ref. 3).
54. Sherman, W.T., Address to the Michigan Military Academy, 19 June 1879.
55. Silverberg, J. & Gray, J.P., *Aggression and Peacefulness in Humans and Other Primates,* Oxford University, 1992.
56. Singer, P., *The Expanding Circle,* Clarendon, Oxford, 1981.
57. Sipes, R.G., "War, sports and aggression," *American Anthropologist,* 75: 64, 1973.
58. Smith, D.J., *The Sleep of Reason,* Century, London, 1994.
59. Smoluchowski, R. et al. (eds), *The Galaxy and the Solar System,* University of Arizona, 1986.
60. Solzhenitsyn, A., *The Gulag Archipelago,* Harper & Row, New York, 1973.
61. Sorokin, P., *Social and Cultural Dynamics,* Sargent, Boston, 1957.
62. Spark, R., "We are not all to blame . . . ," *Mail on Sunday,* 28 November 1993.

363. Stent, G.S. (ed), *Morality as a Biological Phenomenon,* University of California, 1980.
364. Stevenson, J.G., "Song as a reinforcer," in Hinde (Ref. 185).
365. Stevenson, R.L., *The Strange Case of Dr Jekyll and Mr Hyde,* Davos, London, 1886.
366. Stevenson, R.L., *Across the Plains,* Davos, New York, 1906.
367. Stiglmayer, A. (ed), *Mass Rape,* University of Nebraska, 1994.
368. Storr, A., *Human Destructiveness,* Routledge, New York, 1991.
369. *Sunday Times,* London, 31 October 1993.
370. Surin, K., *Theology and the Problem of Evil,* Blackwell, Oxford, 1986.
371. Symons, D., *The Frontier of Human Sexuality,* Oxford University, 1979.
372. Thomas, K., *Religion and the Decline of Magic,* Weidenfeld & Nicolson, London, 1971.
373. Thomas, T.L., "The far look," in Amis & Conquest (Ref. 6).
374. Thornhill, R., "Adaptive female mimicking . . . ," *Science,* 205: 412, 1979.
375. Thornhill, R., "Rape in *Panorpa* scorpionflies," *Animal Behaviour,* 28: 52, 1980.
376. Thornhill, R. & N.W., "Human rape: an evolutionary analysis," *Ethology and Sociobiology,* 4: 137, 1983.
377. Thornhill, R. et al., "The biology of rape," in Tomaselli & Porter (Ref. 384).
378. Tierney, P., *The Highest Altar,* Viking, New York, 1989.
379. Tiger, L. & Fox, M., *The Imperial Animal,* Holt, New York, 1970.
380. *Time* magazine, 6 June 1994.
381. Timms, R., *Doubles in Literary Psychology,* Bowes, Cambridge, 1949.
382. Tinbergen, N., *Curious Naturalists,* Anchor, New York, 1958.
383. Toch, H., *Violent Men,* Aldine, Chicago, 1969.
384. Tomaselli, S. & Porter, R. (eds), *Rape,* Blackwell, Oxford, 1986.
385. Trevor-Roper, H.R., "Witches and witchcraft," *Encounter,* 28: 13, 1967.
386. Trivers, R., "The evolution of reciprocal altruism," *Quarterly Review of Biology,* 46: 35, 1971.
387. Trivers, R., "Parent-offspring conflict," *American Zoologist,* 14: 249, 1974.
388. Trivers, R., *Social Evolution,* Benjamin Cummings, California, 1985.
389. Turnbull, C., *The Forest People,* Jonathan Cape, London, 1961.
390. Turnbull, C., *The Mountain People,* Jonathan Cape, London, 1973.
391. Turney-High, H., *Primitive War,* University of South Carolina, 1971.
392. UNESCO, "The Seville Statement on Violence," *Sixth International Colloquium on Brain and Aggression,* Seville, 1986.

393. Van Beek, W.E.A., "The dirty smith . . . ," *Africa*, 62: 38, 1992.
394. Van Dyne, G. (ed), *The Ecosystem Concept in Natural Resource Management*, Academic, New York, 1969.
395. Van Gogh, V., *The Complete Letters of Vincent Van Gogh*, New York Graphic Society, 1958.
396. Vayda, A., *War in Ecological Perspective*, Plenum, New York, 1976.
397. Von Neumann, J. & Morgenstern, O., *Theory of Games and Economic Behavior*, Princeton University, 1953.
398. Waal, F. de, *Chimpanzee Politics*, Jonathan Cape, London, 1982.
399. Waal, F. de, *Peacemaking Among Primates*, Harvard University, 1989.
400. Waal, F. de, "Food sharing and reciprocal obligations . . . ," *Journal of Human Evolution*, 18: 433, 1989.
401. Waal, F. de, "Aggression as a well-integrated part of primate social relationships," in Silverberg & Gray (Ref. 355).
402. Wallace, B., "Misinformation, fitness and selection," *American Naturalist*, 107: 1, 1973.
403. Walther, F.R., "Flight behaviour and avoidance of predators," *Behaviour*, 34: 184, 1969.
404. Walzer, M., *Just and Unjust Wars*, Scribners, New York, 1977.
405. Ward, D., "Police chief labels two boy killers as evil," *Guardian*, 25 November 1993.
406. Watson, L., *Lifetide*, Hodder & Stoughton, London, 1979.
407. Watson, L., *Whales of the World*, Hutchinson, London, 1981.
408. Watson, L., *Dreams of Dragons*, Sceptre, London, 1987.
409. Weinrich, J.D., "Homosexual behaviour in animals," in Forleo & Pasini (Ref. 127).
410. Westing, A., "War as a human endeavour," *Journal of Peace Research*, 3: 261, 1982.
411. Wheeler, J., *Some Strangeness in the Proportion*, Addison, Reading, 1980.
412. White, G.F., "Media and violence," *Aggressive Behavior*, 15: 423, 1989.
413. Whitehead, A.N., *Process and Reality*, Cambridge University, 1929.
414. Whitmire, D.P., "Periodic mass extinctions . . . ," *Nature*, 308: 713, 1984.
415. Wiberg, H., "What have we learned about peace?," *Journal of Peace Research*, 15: 110, 1981.
416. Wilkinson, A., "Conversations with a killer," *New Yorker*, 18 April 1994.
417. Wilkinson, G.S., "Reciprocal food sharing in the vampire bat," *Nature*, 308: 181, 1984.
418. Williams, G.C., *Adaptation and Natural Selection*, Princeton University, 1966.

419. Williams, G.C., "Mother Nature is a wicked old witch," in Nitecki & Nitecki (Ref. 288).
420. Williams, G.C., "A sociobiological expansion of *Evolution and Ethics*," in Paradis & Williams (Ref. 297).
421. Williams, G.C., *Sex and Evolution*, Princeton University, 1975.
422. Williams, G.C., "Comments on the evolution of schooling," *Publications of the Michigan State University Museum*, 2: 351, 1984.
423. Williams, L., "The feeling of being stared at," *Journal of Parapsychology*, 47: 59, 1983.
424. Wilson, C. & Seaman, D., *Encyclopaedia of Modern Murder*, Pan, London, 1989.
425. Wilson, E.O., *Sociobiology*, Harvard University, 1975.
426. Wilson, E.O., *On Human Nature*, Harvard University, 1978.
427. Wilson, E.O., *Biophilia*, Harvard University, 1984.
428. Wilson, J.Q., *Time* magazine, 23 August 1993.
429. Wilson, M. & Daly, M., "The man who mistook his wife for a chattel," in Barkow (Ref. 27).
430. Wilson, M. & Daly, M., "Competitiveness, risk taking and violence . . . ," *Ethology and Sociobiology*, 6: 59, 1985.
431. Winter, J.M., "Causes of War," in Groebel & Hinde (Ref. 162).
432. Wolf, L.L., "Prostitution behavior in a tropical hummingbird," *Condor*, 77: 140, 1975.
433. Wright, R., *The Moral Animal*, Pantheon, New York, 1994.
434. Yablonsky, L., *The Violent Gang*, Macmillan, New York, 1962.
435. Yeats, W.B., "The second coming," *Collected Poems*, Methuen, London, 1933.
436. Young, D., *Origins of the Sacred*, Little, Brown, New York, 1991.
437. Zahani, A., "Mate selection . . . ," *Journal of Theoretical Biology*, 53: 205, 1975.
438. Zuk, M., "The role of parasites in sexual selection," *Advances in the Study of Behaviour*, 21: 39, 1992.
439. Zweig, C. & Abrams, J. (eds), *Meeting the Shadow*, Tarcher, Los Angeles, 1991.

INDEX

action-aggression addiction, 255
Adaptation and Natural Selection (Williams), 252
Adelie penguins, 54–56
adultery, 108, 120–22, 198
  ritualized, 153
Aeschylus, 167
African Queen, The (film), 271–76
aggression, 123–40, 141–81, 272–74, 280
  among chimpanzees, 128–37
  among hyenas, 59–60
  aversion to, 154–60
  criminal, *see* crime
  crowding and, 35–36
  harmless expressions of, 183–84
  mass, *see* war
  physiological and neurological sources of, 161–63
  ritualized, 141–49, 151–53, 165–69, 184–85, 252
  sexuality and, 123–27, 149–51, 163, 177–80
  verbal, 169
  see also murder
agriculture, development of, 110, 239
AIDS, 114, 118, 279
Akhenaton, 111
Alexander I, Tsar of Russia, 139
Alexander the Great, 173
Alice Through the Looking Glass (Carroll), 23
Amazon, 19, 71, 142, 158
Amin, Idi, 229
Amis, Martin, 216
ammonites, 12
Anatomy Lesson of Dr. Tulp, The (Rembrandt), 92
anchovies, 68
Andersen, Hans Christian, 267–68
Andersson, Malte, 116–17
androgens, 60
animism, 263
Answer to Job (Jung), 269
ants, 19
Apollo astronauts, 16
Apsche, Jack, 228
Aquinas, St. Thomas, 266
Arab-Israeli war, 175
Archimedes, 14
Arendt, Hannah, 289
Aristotle, 7–9, 14, 29, 37, 156, 167, 248, 276
Arnhem Zoo, 128–29, 131–34, 140
Aryan Nations, 280
Asahara, Shoko, 278
Asamanampo, 144
Asmat people, 141, 143–48, 151–53, 156–58, 170, 199, 231
Assyrians, 173
Atahualpa, King of the Incas, 106
Audubon, John James, 36
Augustine, St., 87, 174, 266, 269
Augustus Caesar, 111
Aulton, Renee, 193
Aum Shinrikyo, 278–80
Auschwitz, 286–87
Australia, 256
  aborigines of, 169, 239
  handgun murders in, 190
  rabbits introduced to, 27, 29–31
Avatip people, 155
Axelrod, Robert, 81–83, 87, 201
Aymara Indians, 166
Azande people, 240
Aztecs, 111, 166

baboons, 73–74, 137, 252
Bacon, Francis, 7
bacteria, 18
Baffinland, 169
"bait fish," 68
Baker, Robin, 121–22
Bangladesh, 180
Ba'Mbuti pygmy people, 137–38
Bantu, 235
baobabs, 42–44
barn swallows, 121

Bathory, Elizabeth, 224
bats, vampire, 79–80
battered-woman syndrome, 255
Bayard, 189
bears, 182
Beatles, 98
beauty, standards of, 99–100
bee-eaters, 60, 123
bees, African, 28
beetles, 70
Bell, Mary, 210–11, 215, 216, 267
Belzec, 287
Berkowitz, David, 223
Bernoulli, Jacob, 13
Betzig, Laura, 110–11
Bible, 263, 269
birds
  emergence of flight in, 102
  see also specific species
blackbirds, 69
black holes, 21–22
blacksmithing, 237
blood feuds, 200–201
Bly, Robert, 247
Bogart, Humphrey, 271
Borgia, Cesare, 229
Bosnia, 134
Bounty (ship), 126
bowerbirds, 69
brain
  aggression and structures of, 161, 163
  hemispheres of, 101
Brecht, Bertolt, 167
bricolage, cultural, 157
Britain, 139–40
  murders in, 190, 205–11
Bronze Age, 165
Brothers Karamazov, The (Dostoyevsky), 243
Buddhism, 263
Buenrostro, Dora, 193
Bulger, James, 205–11, 215, 220–23, 232, 289
Bundy, Ted, 218, 228, 232
Burger King, 280–81
Burgess, Anthony, 209
Burkert, Walter, 165
Bush, George, 98
Bushmen, 125, 142, 164, 205, 239
Butler, Samuel, 57
Byrne, Richard, 73

cactus, prickly pear, 31
Caesar, Julius, 98
Caligula, 229
Calley, William, 287–88
Cambodia, 269
Canada, murders in, 190, 197
cannibalism, 141–48, 153, 245
  among animals, 251–52
capitalism, 291
Carib tribe, 159
Carroll, Lewis, 118
Catholic Church, 129
cats, crowding in, 36
CBS Television, 222
chacma baboons, 73–74
chaffinches, xiv
Chagnon, Napoleon, 149–51
chambered nautilus, 11–12
Chamberlain, Neville, 151
Chaplin, Charlie, 98
Cheyney, Dorothy, 258–59
Child's Play III (film), 208–9, 216
Chile, 166, 269
chimpanzees, 52–53, 75, 120, 128–37, 139, 140, 262
China, 179
  ancient, 111, 189
  Great Wall of, 172
chivalry, age of, 189
Christian, Fletcher, 126
Christianity, 127, 153, 166, 242
  right-wing militias and, 280
  witch-hunts and, 244, 245
  see also Judeo-Christian tradition
chromosomes, 49, 108, 162, 176
Churchill, Winston, 199, 229
Church of England, 206
Cilento, Diane, 120
Cincinnati Zoo, 26
Circus Maximus, 183
Clarke, Arthur C., x
Claudius, 111
Clausewitz, Carl von, 170, 171

Clinton, Bill, 98, 247
Clockwork Orange, A (Burgess), 209
cloning, 49
Coast Guard, U.S., 84
Code of the Streets, The, 187
Cohn, Norman, 243–45
Coleridge, Samuel Taylor, 215
Communists, "witch-hunts" against, 246
composite portraits, 99–100
conformity, 212
Congress, U.S., 83
Constitution, U.S., 280
corals, 18–19
Corballis, Michael, 101
Coren, Stanley, 98
corpus callosum, 101
Covenanters, 265
Cretaceous period, 12
crime, violent, 185–87, 274
  genetic component in, 226–27
  see also murder
Croats, 180
crowding, responses to, 35–36, 39
Crozier Mission (New Guinea), 148
cuckoos, 61–63, 66
culture, transmission of, 213–14, 257–59
Cyprus, feuds in, 200
Czechoslovakia, 140

Dachau, 287
Dahmer, Jeffrey, 4, 218, 223, 228, 232
Daly, Martin, 191, 192, 195, 196
Dante, xi, 229
Darius, 173
dark stars, 23
Darwin, Charles, 23, 45, 52, 99, 104, 124, 195, 249, 255–58, 263, 290
Darwinian evolution, xv, 51, 64, 71, 79, 85–87, 110, 158, 174, 177, 198, 213, 252–54
Davies, Paul, 15
Dawkins, Richard, 49, 57, 62, 87, 213
de Waal, Frans, 128–29, 131–34, 136
deer, 252
Dentan, Robert, 156
depression-suicide compulsion, 255
determinism, biological, 254
Devonian period, 12
Dillard, Annie, 249, 281
dinosaurs, extinction of, 22
Dirac, Paul, 15
DNA, 57, 94–95
Donana Reserve, 61
Dorobo people, 235–37
Dostoyevsky, Fyodor, 243
Douglas firs, 32
Downs, Elizabeth, 193
drug addiction, 264
ducks, 123–24, 251
dyslexia, 274

eagles, 59
Easter Island, 175
Eban, Abba, 175
Ebola virus, 278
Egypt, ancient, 111
Ehrlich, Paul, 38
Eibesfeldt, Eible, 142
Eichmann, Adolf, 289
Elizabeth II, Queen of England, 281
electromagnetism, 14–15
elephants, 42–44, 192
elephant seals, 69, 119, 124–25
Emerson, Ralph Waldo, 89
Emlen, Stephen, 123
English Civil War, 265
Equus (Shaffer), 274
European swallows, 62–63
Evans-Pritchard, Edward, 240
Executioner's Song, The (Mailer), 219
extinction, 22–23, 86
eyes of killers, 218–23, 283

Falkland Islands War, 129
Fascism, 38
familicide, 198
family life, 250
Fatal Attraction (film), 198
Federal Bureau of Investigation (FBI), 224
feuds, 200–201
films
  violent, 209
  see also specific films

fish, 251–52
flatid bugs, 70
Ford, Gerald, 98
Fossey, Dian, 73
Fox, Michael, 110
France, 140
Frank, Laurence, 60
Frankenstein (Shelley), 268, 284
Franz I, Emperor of Austria, 139
fratricide, 59, 63, 64
free will, 254, 255
Freud, Sigmund, 104, 123, 163, 195, 266
Friedrich Wilhelm, King of Prussia, 139
frontal lobes, 161
fruitfly, Mediterranean, 28

Gacy, John Wayne, 219, 223, 228–29
Gage, Phineas, 161
Galileo, 263
Galton, Francis, 99
game theory, 78, 81, 86, 275
gangs, 184–85
gay men, 114
geese, 211–12
Gein, Edward, 225
generativity, 102
genes, 49–54, 56–65, 67–70, 72, 75–78, 81, 86–87, 213, 277–78, 284–85, 289–90
  aggression and, 161–63, 175–76
  attractiveness and, 100
  criminality and, 226–27
  handedness and, 97–98
  natural selection and, 255–56, 260
  sacrifice and, 167
  sexuality and, 108–9, 111–13, 117, 123, 179
  view of world centered on, 253
genocide, 220
Germany, 160, 169
  Nazi, 140, 180, 269–70, 286–87, 289
Gerritszoon, Harmen, 91
gibbons, 107, 120
Gibbs, Nancy, 5
Gide, André, 234
Gilmore, Gary, 219
Girard, René, 166
Glover, Stephen, 220
goats, domestic, 37–38
Goethe, Johann Wolfgang von, xi
Goldilocks Effect, 15, 20, 22, 24, 25, 29, 32, 42, 85, 156, 181, 213, 231, 276, 284
Golding, William, 208
Goodall, Jane, 133–34, 262
Goode, William, 191
gorillas, 73, 75, 106
gravity, 14–15
Greece, ancient, 173, 174
Gribbin, John, 15, 21, 22
grizzly bears, 182
ground squirrels, 251
Guardians of American Liberty, 280
Gulf War, 214
guns, 189–90
Gypsies, Nazi extermination of, 286

Hadean period, 17–18
Hadrian's Wall, 172
Haldane, J. B. S., 50, 53
hamadryas baboons, 252
Hamilton, William, 52, 53, 56, 63, 77, 117–18, 253
Hamlet (Shakespeare), 209
Hammurabi, 111
handedness, 95–98
handguns, 189–90
Hansen, Robert C., 223
Hardy, Thomas, 158
hares, snowshoe, 39–40
Hawaii, 28
headhunters, 141–48, 152, 153
Heidnik, "Bishop" Gary, 228
Hepburn, Katharine, 271
herons, 59
Hoess, Rudolf, 286–88
Hick, John, 269
High Wind in Jamaica, A (Hughes), 208
Hindley, Myra, 4
Hitler, Adolf, 140, 286
Hobbes, Thomas, 83
Holocaust, 286, 289
Holy Alliance, 139

Hombre (film), 120
homicide, *see* murders
homosexuality, 114, 251
Hope Bay (Antarctica), 54–56
Hopi Indians, 169
hormone-mediated idiopathic hypoglycemia, 39
hormones, 60
  aggression and, 125–26, 162
horses, violence against, 274
Hoyt, Waneta, 193
Hrdy, Sarah, 250
Hubble time, 14
Hudson Bay Company, 39
Hughes, Richard, 208, 217
hummingbirds, 113
Humphrey, Nicholas, 75, 91–93
Huns, 173
Hunt, Morton, 185–86
hunter-gatherers, 235, 239
Hurrians, 173
Hussein, Saddam, 214
Hutu people, 220
Huxley, Thomas Henry, 249–50, 252, 253, 255, 259, 277, 283
hyenas, 58–60
Hyksos, 173
hypergyny, 112
hypothalamus, 161

Ik people, 237
Incas, 106, 166–68
incest, 123, 167
inclusive fitness, 63, 67, 99
Indecent Proposal (film), 113
India, 75
  ancient, 111
infanticide, 63–64, 76, 120, 181, 192–95, 199
  cannibalistic, 245
  female, 150
innocence, loss of, 208
Inquisition, 247
intelligence, social, 74–75
Inuit people, 34, 169
Ionesco, Eugene, 167
Iraq, 214
Irenaeus, Bishop of Lyons, 269
Irian, 141, 142, 144, 152, 153
iridium, 22
Iron 56, 20
Islam, 214, 242, 263
Israeli *kibbutzim*, 123
Ituri forests, 137

Jack the Ripper, 224
jackdaws, 211
Japan, 140, 167, 179, 258
  feudal, 152
  handgun murders in, 190
  terrorism in, 278–79
Jefferson, Thomas, 9
Jews, Nazi extermination of, 270, 286–87
Jivaro people, 152
Joan of Arc, 244
Journal of Peace Research, 164
Joyce, James, ix
Judeo-Christian tradition, 7, 38, 97, 102, 166–67, 242, 263, 269
Julius Caesar, 111, 230
Jung, Carl Gustav, 247–48, 257, 266, 268, 269, 282
Justinian, 162
Juvenal, 1

Kalahari Desert, 67, 142, 205, 239
Kamchatka, 41
Kaplan, Robert, 272
Kapsiki people, 237
Kassites, 173
Katz, Fred, 286
Keegan, John, 172
Kelly, Walt, 248
kelps, 40–42
kemari (game), 86
Kennedy, John F., 229
Kenya, 236
kibbutzim, 123
Kierkegaard, So/ren, 166, 168
Kikuyu people, 167, 236
"killer" bees, 28
killer whales, xii–xiii
  false, 84–85
kinship, 49, 52–54, 80, 87, 149–50
  sexuality and, 108

Kipsigis people, 236
Kissinger, Henry, 110
Kivi, Lynn, 66
Klein, Melanie, 280
Koran, 263
Kubrick, Stanley, 209
Kummer, Hans, 212
!Kung Bushmen, 142, 164

language, 261
  origin of, 214–15
langurs, 75–77, 250
Large Numbers, Lore of, 14–15
League of Nations, 86
Le Boeuf, Burney, 119, 124
lemmings, 33–35
Leonard, Elmore, 120
Leonardo da Vinci, 98
leopard seals, 55
Levi, Primo, 285
Lévi-Strauss, Claude, 157–58
life
  origin of, 16–19
  of stars, 20–21
limbic system, 161
lions, 76, 77
Lobedu people, 240–41
London Zoo, 26, 58
Lord of the Flies (Golding), 208
Lorenz, Konrad, 138, 163
Lucas, Henry Lee, 223
lynxes, 39

Maasai people, 118–19, 235, 236
macaques, 176, 258
Macbeth (Shakespeare), 193
Machiavelli, Niccolò, 74–75, 131
Mafia, 134
magic, 237–39
magpies, 61–62
Mailer, Norman, 219
male bonding, 135–36
mallard ducks, 251
Manson, Charles, 218, 220, 232
Maoris, 171–72
Mapuche Indians, 166
Marines, U.S., 287–88
Maring people, 171
Marquesas Islands, 28
Marsh, Peter, 183, 185
marsh wrens, 69
marsupials, 49, 256
Marxism, 50
mass murderers, *see* serial killers
Masters, Brian, 228
matricide, 196
Mayans, 166
McCarthy, Joseph, 246–47
McEwan, Ian, 208
McManus, Christopher, 93
McVeigh, Timothy, 279
Medea Syndrome, 199
Mediterranean fruitfly, 28
Mehinaku tribe, 159, 160
memes, 213–14, 225, 232, 261, 274
  of monkeys, 258
Mendel, Gregor, 52
Michigan Militia Corps, 280
Midgley, Mary, 290
Midianites, 180
Milgram, Stanley, 288
militias, right-wing, 279–80
Milky Way, 13, 23
Mill, John Stuart, 276–77
Milton, John, xi
monkeys, 137, 176, 258
monogamy, 107, 120, 121, 127
Montaigne, 97
Montezuma, 111
Moore, Demi, 113
Moore, Omar Khayyam, 238–39
Moors, 167
Mormons, 107
morphic fields, 256–59
Morris, Desmond, 26
Moulay Ismail, Emperor of Morocco, 197
Mountain People (Turnbull), 237
Mugging, The (Hunt), 186
Mundurucú people, 142, 143
Munich pact, 151
murders, 126, 189–211, 215–31
  of children, 205–11, 215, 220–23 (*see also* infanticide)
  defenses of those accused of, 254
  domestic, 124, 191

mass, 286–89
multiple, *see* serial killers
and Other Look, 229–31
ritualized, *see* headhunters
by terrorists, 278–79
mussels, zebra, 28
Mussolini, Benito, 140
My Lai massacre, 287–89

Namaqualand, 285
Nanking, Rape of, 179–80
Napoleon, Emperor of the French, 139
Narses, 162
National Park Service, 84
native Americans, 238
see also specific tribes
natural selection, 252–56, 259–61, 264, 281, 289, 292
nautilus, 11–12
Nazis, 140, 180, 269–70, 286–87, 289
Nefertiti, 111
Nero, 111, 230
neutrinos, 21
Newman, Paul, 120
Newton, Isaac, 263
New Zealand, 29, 171
Nicomachean Ethics (Aristotle), 7
Nijmegen University Hospital, 162
Nilsen, Dennis, 4, 218, 228
Nishida, Toshisada, 130
Norris, Joel, 225–26
Northern Ireland, 128
Norway, lemmings in, 33–35
nuclear time, 14
nuclear weapons, 173–74
Nuremberg Trials, 269–70

Oates, Joyce Carol, 229
Oedipal theory, 195
Offa's Dyke, 172
Oklahoma City bombing, 279
Onesmanam, 144
Oort Cloud, 23
opossums, 49
orchids, 72
Ordovician period, 12
original sin, 87
Othello (Shakespeare), 215
Otterbein, Keith, 152
otters, sea, 41–42
Oxford University, 250

pair-bonding, 107, 109
Pakistan, 180
paleo-psychology, 92
papisj (ritual exchange of wives), 153
paranoia, 280
parasites, 249
passenger pigeons, 36
Pathics, Principles of, 30–38, 42, 57, 124, 152
pathology, 30
patricide, 196
Pausanias, 165
pelicans, 65–66
penguins, 54–56
Persian empire, 173
Peru, 166
Petrarch, 229
Philippe, Joseph, 224
phytoplankton, 19
Piaget, Jean, 129
pigeons, passenger, 36
Pilgrims, 28
Pitcairn Island, 126–27
placental mammals, 49
plagues, 33
plantains, 28
planthoppers, 71
Pogo (comic strip), 248
polygamy, 106–7
polygyny, 106, 112–13
Polynesians, 126–27, 175
portrait painting, 91–95
postpartum depression, 193–94
post-traumatic stress disorder, 255
Potato-Eaters, The (Van Gogh), 93
Potts, Frank, 223
pregnancy, human, 48–49, 108
artificial, 202
from rape, 179
miscarriages in, 64–65, 125
Presbyterianism, 265
prickly pear cactus, 31
Primitive War (Turney-High), 170
Prisoner's Dilemma, 78–79, 81–82

progestin, 125
pronking, 67–68
prosocial acts, 212
prostitutes, 112–15
  murders of, 224
pseudospeciation, 142
Psycho (film), 225
psychoanalytic theory, 245
Ptolemy, 229

rabbits, 27, 29–31
Raine, Adrian, 163, 227
Rais, Gilles de, 224
Ramirez, Richard, 232
rape, 177–80
  in animals, 251
Rasputin, Grigory, 229
rats, crowding in, 36
Read, Piers Paul, 210
Reagan, Ronald, 98
Redford, Robert, 113
red giants, 20
redwoods, 32
religion, 213–14, 262–65
  see also Christianity; Judeo-Christian tradition
Rembrandt, 91–94, 102
revenge, 199–201
Ridley, Mark, 105, 213
Robie, Derrick, 221, 223
Romans, ancient, 97, 111–12, 172, 174, 183
Rubin, Lillian, 131
Rule, Ann, 228
Russia, 139–40
Ruth, Babe, 98
Rwanda, 5–6, 75, 127, 128, 220, 269, 273–74

Sachsenhausen, 287
sacrifice, 165–68
salamanders, 251
Samburu people, 235
San Bushmen, 125
Sand Reckoner, The (Archimedes), 14
sardines, 68
Satanic cults, 283–84
scapegoating, 237, 283
Schwartz, Anne, 218, 228
scorpion-flies, 72, 177, 178
Seafarth, Robert, 258–59
seals
  elephant, 69, 119, 124–25
  leopard, 55
sea otters, 41–42
sea urchins, 41–42
Sebei people, 201
Self, Will, 218
Semai people, 154–59, 164
Senate, U.S., 83, 246
sequoia, giant, 32
Serbs, 180, 269, 273
serial killers, 218–19, 224–26, 228–29, 232
serotonin, 162
Seven Deadly Sins, 7, 153
Seville Statement on Violence, 138
sexuality, human, 104–115, 117–27, 140
  aggression and, 123–27, 149–51, 163, 177–80, 196–98
  ritual and, 153
sexual reproduction, 49, 105
Seychellois people, 157
shadow, Jungian, 247–48, 268, 282–83
Shaffer, Peter, 274
Shaka, Chief of Zulus, 170–71
Shaw, George Bernard, ix, 253–54
Sheldrake, Rupert, 256–59
Shelley, Mary, 268
Sherman, Gen. William Tecumseh, 174
sibling rivalry, 59
Silence of the Lambs, The (film), 225
Singer, Peter, 260
Smith, David, 220
Smith, Eric, 221–24, 229, 232, 267, 283
Smith, John Maynard, 78
Smith, Susan, 192–93, 199
snowshoe hares, 39–40
social intelligence, 74–75
socialism, 291
sociobiology, 56
Solzhenitsyn, Aleksandr, 289
Somalia, 128, 134, 201, 273
sperm competition, 121–22
spiders, 70, 251
Spielberg, Steven, 167

spirals, 11–13
"spontaneous" abortion, 64–65
sports
  murder and, 204
  ritualized aggression in, 147, 148
springbucks, 67
Springsteen, Bruce, 202
squirrels, 251
Stalin, Joseph, 140
status, 271
stepparents, 194–95
Stevenson, Robert Louis, xi, 265–67
stilts, 50–51
stink bugs, 70
Stone Age, 172
Stonehenge, 165
Strange Case of Dr Jekyll and Mr Hyde, The (Stevenson), 265–68, 284
suicide, 34–35, 64
Suk people, 235
supernovas, 20–21
Surinam, 134
swallows, 62–63, 121
Sweden, handgun murders in, 190
symmetry, 94–96, 99

tamarisks, 28
T'ang dynasty, 111
Tanzania, 236
Tartars, 173
Ten Commandments, 7
territoriality, 107, 142, 182, 184
terrorists, 168, 278–80
testosterone, 60, 126, 162
Texas Chainsaw Massacre, The (film), 225
thalamus, 163
Thermodynamics, Second Law of, 16
Thomas, Theodore, 223
Thompson, Robert, 206–11, 215–18, 220–22, 229, 232, 267, 283
Thornhill, Randy, 178
Tiberius, 111
Tiger, Lionel, 110
time, measurement of, 14
Tit for Tat, 82–83, 85, 187, 200, 275–77
Torquemada, Tomás de, 229
Tory Island, 185
Treblinka, 287
trench warfare, 86
Triple Entente, 140
Trivers, Robert, 80–81, 250, 267
tropical rain forests, 19
Turkana people, 235
Turkey, feuds in, 200
Turks, 173
Turnbull, Colin, 137–38, 237
Turney-High, Harry, 170, 171
Tutsi people, 220
"Twinkie Defense," 254
twin studies of criminality, 227

Udayama, 111
Uganda, 201, 236, 269
UNESCO, 167
United Nations, 279
universe
  age of, 14
  and black holes, 21–22
  origins of, 24
utilitarianism, 276

vampire bats, 79–80
van Gogh, Vincent, 93–94
Venables, Jon, 206–10, 217–18, 220
vervet monkeys, 258–59
Vietnam War, 190, 269, 287–88
violence, *see* aggression
Volterra, Vito, 38
Volterran principle, 39–40

Wallace, Henry Louis, 223
Wa Ndorobo, *see* Dorobo people
war, 128, 134, 138, 160–61, 164, 169–76, 181
  mass murder during, 286–89
  rape during, 179–80
  religious, 214
  weapons in, 188–89
wasps, 70, 72, 251
Waura tribe, 160
weaning, 251
whales, xii–xiii, 84–85, 192
Wheeler, John, 16
Whitehead, Alfred North, xvi

Whiten, Andrew, 73
white supremacy, 280
Whitman, Charles, 203
widow birds, 115–17
Wilberforce, Bishop, 249
Wilkinson, Alec, 219, 228–29
Williams, George, 252–53, 259, 260, 275, 281
Wilson, Alexander, 36
Wilson, Edward, 56–57, 252
Wilson, James, 186
Wilson, Margo, 191, 192, 195, 196
witches, 237, 239–46, 248
Wolf, Larry, 113
Woodhenge, 165
World War I, 86
World War II, 170, 179, 180, 188, 287
wrens, marsh, 69
Wright, Robert, 261, 262, 264, 271, 277

xenophobia, 140
Xinguanos, 158–60, 164–65, 168

Yakuzas, 167
Yanomaö people, 149–51, 157–58, 170, 171, 179, 188
Yeats, William Butler, ix
yellow fever epidemics, 32–33

zebra mussels, 28
Zulu people, 170–71

9 780060 927905